MACRODYNAMICS

THE ARNOLD AND CAROLINE ROSE BOOK SERIES
OF THE AMERICAN SOCIOLOGICAL ASSOCIATION

The Editor and Editorial Board gratefully acknowledge the contribution of Paul Colomy, University of Denver, and Robert Alun Jones, University of Illinois—Urbana, as expert reviewers of this manuscript.

MACRODYNAMICS

Toward a Theory on the Organization of Human Populations

JONATHAN H. TURNER

RUTGERS UNIVERSITY PRESS
New Brunswick, New Jersey

Library of Congress Cataloging-in-Publication Data

Turner, Jonathan H.
Macrodynamics : toward a theory on the organization of human populations / by Jonathan H. Turner.
p. cm. — (Arnold and Caroline Rose Book series of the American Sociological Association)
Includes bibliographical references (p.) and index.
ISBN 0-8135-2162-9
1. Macrosociology. I. Title. II. Series.
HM24.T838 1995
301′.01—dc20 94-40290
CIP

British Cataloging-in-Publication information available

Published by Rutgers University Press, New Brunswick, New Jersey
Manufactured in the United States of America

To the Memory of
My Mother, Marie Turner

CONTENTS

PREFACE

THIS BOOK SEEKS to develop abstract concepts, models, and principles that can explain the organization and dissolution of populations in space and over time. Like my other efforts to develop theory, I visualize my job as presenting abstract theoretical ideas that might be of interest to my fellow theorists and to empirical researchers. True, at first blush, the theoretical ideas in these pages will seem abstract and inadequate to the nuances of empirical situations. Yet this is what theory must do: it must ignore the particular and contingent; it must seek the universal and invariant; and it must sustain a level of abstraction above the very empirical details that excite researchers. And so, it will be necessary to connect the abstractions in this book to the particulars of historical or present-day cases. This necessity should not be used to mount a critique, as is so often done in these "postmodern" and "postpositivistic" times, but to engage in an exciting and important intellectual exercise.

I find it sad that sociology has become so differentiated and compartmentalized into disparate research traditions, paradigms, and metatheories that integration of the field by general theory now seems impossible. Each division sustains its own particularistic symbols and often reacts negatively to those of other camps. This state of disintegration can be explained by application of the macro-level principles developed in chapter 8; for as those principles underscore, once a population begins to differentiate into particularistic subcultures, with little consensus over generalized symbols and weak centers of regulatory power, such as the American Sociological Association, the process of breakdown is difficult to arrest. Sociology is in the midst of this process, and I am afraid that little can be done to stop its breakdown into a series of disconnected subfields where general theory is considered just another isolated camp with its own irrelevant symbols. In contrast to this trend, my hope is that general theories can provide a basis for some symbolic unity, although this hope is perhaps only wishful thinking.

Yet as I seek to demonstrate in this book, what are usually considered incompatible approaches—functional, evolutionary, bioecological, and conflict theories—are highly compatible. Each can add something to the others in producing a more complete explanation of macro-level phenomenon. Thus, by producing a theory that cuts across paradigms, perhaps the generalized symbols of abstract theory can bridge some of the partitions that divide sociology. I say this with more hope than conviction, but we must begin to arrest the splintering of the sociological community into so many diverse intellectual factions.

ACKNOWLEDGMENTS

I WANT TO THANK the Academic Senate of the University of California at Riverside for supporting this research and, in fact, all of my research over the last twenty-five years.

I especially want to thank my colleague, Randall Collins, who made invaluable criticisms and suggestions. In addition, two reviewers, Paul Colomy and Robert Alun Jones, also provided very useful suggestions for revising the manuscript. Mark Mizruchi from the Rose editorial board supplied the last round of useful suggestions; my thanks to him as well.

ABSTRACT

THIS BOOK SEEKS to develop a general theory of those generic and fundamental forces that determine the organization of human populations. The most fundamental of these forces are (1) the size and rate of growth of a population, (2) the level of production, (3) the volume and velocity of distribution, and (4) the consolidation and centralization of power. Flowing out of these four primary forces, but also having reverse causal effects on them, are (5) spatial forces, (6) differentiating forces, and (7) disintegrating forces. In turn, sets of subforces, such as inequality and stratification, emerge from the operation of these seven basic forces; and like virtually all dimensions of the social universe, these subforces have reverse causal effects on the very forces that produce them. Thus, the argument of the book is that a relatively small set of forces, and derivative subforces, determines how populations are organized. A theory of macrodynamics must, therefore, explain the operation of this set of forces and subforces.

These forces obviously interact in complex ways, and so, their effects on each other must be analyzed highlighting paths of direct, indirect, and reverse causality. Two strategies are employed to capture, and yet simplify, the complexity of the intersections among these forces: analytical models in which important causal paths are visually represented in two-dimensional space; and propositions in which the key causal connections are stated verbally and, in the appendix, mathematically.

In seeking to develop general models and propositions on these forces and subforces, a synthesis among what are often considered to be incompatible theories—functional theory, conflict theory, human ecological theory, and evolutionary-developmental theory—is performed. The intent of this synthesis is to extract the useful portions of these four theoretical traditions and, then, to bring them together in a more robust and yet parsimonious theory. In the appendix, this goal is pursued to its ultimate conclusion by using the models and propositions to develop ten laws of macrodynamics.

As is perhaps obvious, this book is executed with a view that sociology can, and should, be a "hard science." The goal of all hard science is to explain the universe in terms of abstract models and principles. Sociology can be one of these sciences, as the models and propositions in this book demonstrate.

MACRODYNAMICS

1

MACRODYNAMICS

HISTORY IS THE DUSTBIN of populations that have become assembled and organized, only to fall apart and disintegrate. We need not posit a cyclical view of history to emphasize the point that the structural and cultural organization of a population will eventually prove inviable, opening a society to external conquest or internal collapse. The dynamics of these processes are, of course, the subject matter of history and historical sociology, as well as sociological theory from its very beginnings. It is now time to begin pulling the insights from all these empirical and conceptual efforts into a more unified theory of macrodynamics.

What Is Macro Sociology?

Macro sociology is the study of how relatively large populations of humans are organized over significantly long periods of time in comparatively large amounts of space. In order to constitute the subject matter of macro sociology, however, it is not clear just how large a population must be, how long it must remain a coherent and identifiable whole, and how much space it must use and control. The goal of macro sociology is, of course, to explain such variations in the organization of populations over time and in space, but the question remains: When is analysis macro? I have no firm answer to this question, except to say that macro-level theory focuses on (*a*) populations that number into the thousands, and usually many more; (*b*) populations that remain organizationally and culturally coherent in their environment for at least several decades, and typically many more; and (*c*) populations that control at least a few thousand square miles of territory, and generally much more.

The boundary between macro analysis, on the one side, and meso-level or even micro-level analysis, on the other, is thus a bit vague; but in a connotative sense, it is clear what is involved: macro analysis concerns populations organized as *societal* and *intersocietal* systems; and as such, it is concerned with the varying types of social structures, social categories, and systems of symbols that organize a population over time and in space. In terms of basic types of societies that have existed in human history, small hunting and gathering bands are not the topic of macro analysis, unless these bands are part of a larger confederation organizing many bands over time and in space; the same is true of simple horticultural populations. When we move to more advanced horticultural populations, involving larger populations (several hundred thousand) in larger amounts of territory (many thousands of square miles) and existing over longer periods of time (a few centuries), we are in the domain of macro analysis. The same is true for more advanced societal types, such as agrarian, industrial, and postindustrial.

What Are Dynamics?

Having at least staked out a rough definition of "macro," let me now turn to "dynamics" in order to complete the definition of the subject matter of this book: macrodynamics. The social universe is constructed from processes, or "forces" that drive behavior, interaction, and organization of humans. Since Comte (1830–42), sociology has had an unhealthy tendency to separate "statics" from "dynamics," or in a more contemporary vocabulary, "structure" from "process" and "stability" from "change." For me, everything in the social universe is process, and hence, it is not wise to separate "structure" from the forces that create, sustain, or change such structures. For theories of human social organization must contain conceptualizations of "forces" that, depending on their values and interactions, create and sustain structures and systems of cultural symbols for a time or, alternatively, transform them. A macro-level theory is, *at one and the same time,* one about stability and change or structure and processes. We do not have two types of theories—those on stasis, and another set on change—but one type of theory which seeks to denote the key forces behind human organization and its transformation.

What do I mean by "forces"? A macro-level force is a property of a population that determines the nature of its organization in space and over time. I draw a clear analogy (although for me, it is a homology) to the "hard" sciences when visualizing the macro-level social universe as composed of forces. For just as gravity is a force influencing (along with other forces) the organization of the physical and biotic (and social as well) universe, so there are key features of large-scale human organization that account for how human populations are structured and restructured. It is these forces, and their intersections and relations, that are to be the subject of a macro theory. For they are what explain theoretically the way in which a population is structured over time and in space.

Theoretical explanation in terms of abstract models and principles about basic forces is not, I should emphasize, the only way to explain social reality. There are alternative modes of explanation. The most prominent in macro-level analysis is historical, whereby a phenomenon of interest—say, a revolution or a pattern of conquest—is understood in terms of a descriptive scenario about causal sequences among specific empirical/historical events. For those who prefer this mode of explanation, social reality seems contingent and historical. And as such, it seems not to be amenable to explanations in terms of timeless, generic, and universal laws about a few fundamental forces. To those who advocate this position, I can only say that *all* empirical reality—the physical, organic, and social—is contingent and historical, but this fact does not stop the other sciences from generating explanatory models and principles. The empirical analysis of contingency and history are useful activities in any science; I have no quarrel with the utility of such descriptions. Indeed, they are necessary if we are to know the values and loadings of the forces specified in our theories as we use them to explain particular empirical cases. Moreover, they provide the data base for inductive theorizing as well as for deductive testing of theories. But these empirical variations for all their seeming uniqueness are still generated, I argue, by the same set of basic forces.

The reason for many sociologists' conviction that the social universe is indeterminate—that is, contingent and historical—is this: we must typically work in natural settings in which we cannot know the values of key forces, nor their complex interaction effects. But such is often the case in science, say, for example, among geologists trying to understand earthquakes or meteorologists attempting to predict the weather. Few would argue that these phenomena cannot be studied scientifically. In fact, the basic forces generating the phenomena in question are well known; it is the empirical values and interactions in a particular situation that are not always fully understood. Much macro-level theory has a similar problem: the values of, and interactions among, basic forces are not completely discernible in a robust, natural setting. This fact does not mean, however, that we cannot explain these phenomena theoretically; it only means that our predictions based on a theory will be far from perfect because we are unsure of the values and loadings of the forces, and their interaction effects, in our models and propositions.

Even if one is still convinced that the dynamics of the social universe are highly contingent, if not idiosyncratic, the models and propositions offered in these pages can be used as a framework for interpreting empirical or historical cases. It is not necessary, I believe, to "buy into" my views on these matters, any more than I accept the antiscience epistemology of the many historians whose works have been a constant "reality check" for me as I developed this theory. Let us suspend our respective biases on this issue of contingency and see what we can learn from one another.

Thus, the theory to be developed in these pages revolves around (1) isolating the basic forces organizing populations in space and over time, (b) developing conceptualizations that capture the essential properties of these forces, and (c) developing models and propositions that explain variations in, as well as intersections among, these forces. In pursuing this strategy, the explanations generated can seem empty and devoid of the texture and richness of historical explanations or thick empirical descriptions of events. Yet for me, explanations in terms of abstract models and principles allows us to see what seemingly diverse empirical contexts have in common and to appreciate that descriptions of the empirical flow of events are governed by a small set of basic forces. This ability to "see the forest for the trees" is what makes general theorizing exciting and important. And while it is a very different kind of intellectual activity than historical explanations or purely empirical descriptions, the differences between theoretical and historical explanations need not be seen as a source of incompatibility. Macro-level sociological theory cannot be developed without historical descriptions and explanations; and I hope that theories such as the one to be developed here, for all its flaws, can provide guidelines for how historical descriptions are conducted.

Having said this, the question then becomes: What are the dynamic forces that explain the organization of human populations? The answer will seem familiar and simple, but we should not be snobs because we have long-term familiarity with what shapes our universe. At the most general level, I see the macro-level social universe as being driven by: (1) the size of a population and, relatedly, its rate of growth, its diversity, and its movements; (2) the level of production, or the extensiveness of resource extraction and conversion; (3) the level of distribution, or (*a*) the level of movement of people, materials, and information among the members of a population and (*b*) the scale, scope, volume, and velocity of exchange transactions among the members of a population; (4) the level of consolidated and centralized power, or the extent to which capacities for control and regulation have, respectively, been mobilized and concentrated; (5) the size and configurations of territories, and patterns of settlement within territories; (6) the level of differentiation within and among structural units, social categories, and systems of symbols organizing the members of a population; and (7) the ratio of disintegrative to integrative pressures, or the degree to which (1) through (6) sustain a population in its environment or, ultimately, break it apart.

Some of these forces are part of the definition of macro-level reality, but they nonetheless constitute forces in their own right. Population size and size of territory are, for example, the essential elements in my definition of the macro-level universe, while at the same time they are forces that dramatically affect the values for the other forces and, hence, the way a population is organized. Thus, the apparent tautology is simply part of social reality.

What I am asserting in this list of fundamental forces is this: the size of a population, its productive capacities, its distributive facilities, its consolidation and centralization of power, its size of territory and internal settlement patterns, its overall level of differentiation, and its propensity to fall apart or remain a coherent whole determine how a population is organized. A theory of macrodynamics involves, therefore, a conceptualization of the variable nature of these forces and their complex interrelations.

Figure 1.1 is a skeletal model of the forces that will occupy my attention in this book. Considerably more detail on subforces will be added, and the focus of analysis will constantly shift, but this model can provide a general orientation and, at the same time, allow me to highlight a number of features of such models, which are at the core of the theoretical strategy I will pursue.

In the model, processes flow from left to right, and then back via reverse causal arrows (Stinchcombe 1968). In presenting and positioning the forces as variables and then drawing causal links among them, I am asserting that these are the crucial interconnections. The fact that so many of the causal arrows are reversed underscores the recursive nature of social reality: that is, the outcomes of forces at one point in time feed back and affect their values at subsequent points in time. The real world, therefore, does not look like a statistical path model, in which causality tends to flow in one direction in an effort to explain variance at a single point in time. For example, increases in size of population cause (note signs on arrows) an increase in the level of consolidated/centralized power (as we will see in later chapters, there are other subforces not specified in this skeletal model that are also involved) up to a certain point, and then very large populations begin to decrease the consolidation and centralization of power (again, through additional subforces that will be introduced later). The relationship is thus positively curvilinear, increasing and then decreasing, and is denoted by the sign (+/–). Other basic types of signs to be used in later chapters should be introduced here: (+) = positive relation; (–) = negative relation; (–/+) = negatively curvilinear (i.e., decreasing and then increasing); (=/+) = lagged positive

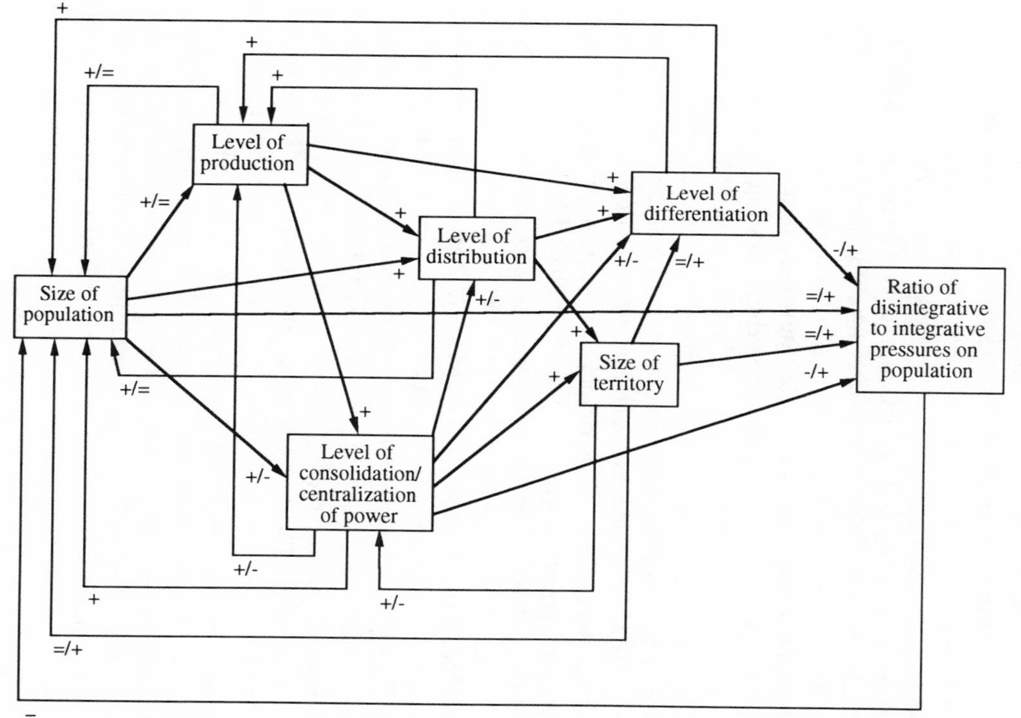

Figure 1.1. A Skeletal Overview of Macrodynamic Forces

(not increasing until a threshold is reached); (=/–) = lagged negative (not decreasing until a threshold is reached); (+/=) = positive relation that levels off (increasing and then stabilizing); (–/=) = negative relation that levels off (decreasing and then stabilizing). In the model presented in figure 1.1, only some of these relations are evident—(+), (+/=), (–/+), and (+/–)—but all of these signs will appear in various models to be drawn.

Let me assert a necessary caveat here to avoid any confusion: the fact that size of population is the first force denoted on the left in figure 1.1 and is the first force discussed in chapter 2 does *not* mean that I advocate some version of "demographic determinism." In a long-term evolutionary sense, the growth of hunting and gathering populations initiated other macro-level forces, but these forces soon came to have effects independent of whatever set them into motion and, moreover, began to have reverse causal effects on population dynamics. Thus, the critical point is this: No one force is paramount. I begin with population because one has to begin somewhere. I could just have easily begun with any other force listed in figure 1.1.

The model presented in figure 1.1 may look complex, but it is one of the simplest that will be presented. In fact, it is only a rough schematic that overviews the book; as such, it is not an explanatory model at all. For this reason, it is simple but the more explanatory models will be considerably more complex; and as this occurs, some might ask: Is this empirical/historical description masquerading as abstract theory? This is a reasonable question, and my answer is this: the forces denoted in the models are not tied to *any* specific empirical context; they are consistently abstract, denoting what I see as generic and universal forces of the social universe. The models become complex because I want to address the ways that many forces intersect with one another—directly and indirectly as well as through reverse causal processes. This desire on my part is not intended to be description but explanation. But there is a problem with the models I will present: many become too complex for easy empirical tests, and I might add, many will try readers' patience. But if one carefully follows the causal paths, as they directly and indirectly connect basic social forces, they offer a sense for the way these social forces affect one another; and in so doing, perhaps what initially seems to be a tedious exercise can become a source of excitement in visualizing how the social universe is constructed. At the very least, one can follow a particular set of causal paths that are of interest. Indeed, any particular path or set of paths can be subject to empirical assessment, becoming in essence a general hypothesis. And when thinking of the causal paths in this way, the other theoretical tactic in my approach comes into play: formulating propositions.

Like an analytical model, propositions specify relationships among forces conceptualized as variables, and as we will see, I tend to construct them from significant causal paths in analytical models. Increasingly, I formulate propositions as "verbal equations" in the sense that variation in one force is seen as a "function" of others. For example, in figure 1.1, I might stress that the level of consolidated power is an additive function of the effects of the size of a population, the level of production and distribution, and the size of territory. Systems of propositions, like the causal paths in analytical models, can also become too complicated. Hence, special care must be taken in formulating propositions in order to capture, on the one hand, some of the complexity of reality and, on the other, to specify testable relations.

Theoretical Eclecticism

In moving toward a theory of macrodynamics, I will be eclectic, borrowing insights from four broad theoretical traditions: functional theorizing, conflict theorizing, ecological theorizing, and evolutionary/developmental theorizing. Although these approaches may seem incompatible, each provides key ideas for a more general theory.

In employing these diverse approaches, my goal is to move sociology further along the path *toward* a theory on the organization of human populations. I take seriously the "toward," because my goal is to lay out the crucial forces and their dynamic relations that will be part of a general theory. Depending on how various scholars prefer to state their theories, the models and propositions presented in chapters 2 to 8 can be useful first steps. The content of the chapters, then, is not a complete theory, but my suggestions for what needs to be part of a general theory. In a long appendix, I present my way of going beyond these first steps toward a theory, but I have resisted incorporating the contents of this appendix into the body of the book. It is there for those who are interested; and it represents my best effort to move further along the path for a general theory. In this appendix, I pull together the suggestions delineated in the chapters and formulate ten laws of macrodynamics. Those who do not believe that scientific theory is possible will find this effort pretentious and preposterous, whereas those who share my commitments to scientific sociology may roll their eyes and view my effort as, to be kind, premature. Nonetheless, I made the effort, whatever its merits. Now, let me begin discussing the preliminary conceptualizations of forces that led me to attempt this effort.

2

POPULATION DYNAMICS

SOME OF THE FIRST sociologists, especially Herbert Spencer ([1874–96] 1898) and Émile Durkheim ([1893] 1933) stressed the significance of population size on the structure of a society, but with a few exceptions (e.g., Goldstone 1990; Carneiro 1967, 1970), the analysis of population dynamics has been relegated to demography, a field somewhat out of the sociological mainstream. It is useful, I think, to bring population as a force of macro-level social organization back into the core of sociological theory. For as we will come to appreciate ever more fully with each successive chapter, the size, composition, distribution, and movement of populations have profound consequences for all other macrodynamic forces.

In opening my analysis with a short chapter on population dynamics, the intent is to emphasize the significance of the absolute number of people and the rate of population growth for other macrodynamic forces. I am not proposing a new demographic determinism, because the very forces that population size and growth set into motion have reverse causal effects on the demography of a population and, eventually, become more significant causal forces than population processes. I need an entry point into the theory; and population dynamics are a good place to begin.

Population Growth and Macrodynamics

The Malthusian Specter

Thomas Malthus ([1798] 1926) was the first to recognize fully the basic relationship between the size and productive capacities of a population. He argued, in essence, that there is an equilibrium point between a population's size and its ability to sustain members; and once this ability is surpassed, death rates increase and the size of a population declines until productive capacities and population size are in balance. This relationship, Malthus felt, is affected by the amount of productive output that can be retained by the general population, and in turn, this amount is affected by a number of other forces: (*a*) the level of technology, (*b*) the level of available natural resources, (*c*) the rates of usurpation by centers of political power of productive surplus, (*d*) the competitiveness of markets, and implicitly, (*e*) the normative standard of living (Lee 1986) influencing how much each member consumes and, hence, what is available for consumption by others.

Thus, Malthus recognized that the size of the population is related to the forces of production, power, and distribution in a society, as well as to culturally defined standards of living, but ultimately there is what I call the "Malthusian base," or the biological capacity to sustain a given number of people. Ronald Lee (1986) prefers to conceptualize this base as a "space" within which population size can fluctuate, but the essential point is the same: patterns of social organization are ultimately constrained by the realities of biologically supporting the population. But does overpopulation initiate Malthus's "four horsemen"—war, disease, pestilence, and famine—and increase death rates until an equilibrium point is reached? Historically, the answer to this question has varied, but one response to a growing population has been technological innovation and expanded production in order to ward off the four horsemen. Indeed, the sociocultural elaboration of society into ever more complex forms was historically initiated by efforts to stay above the Malthusian base.

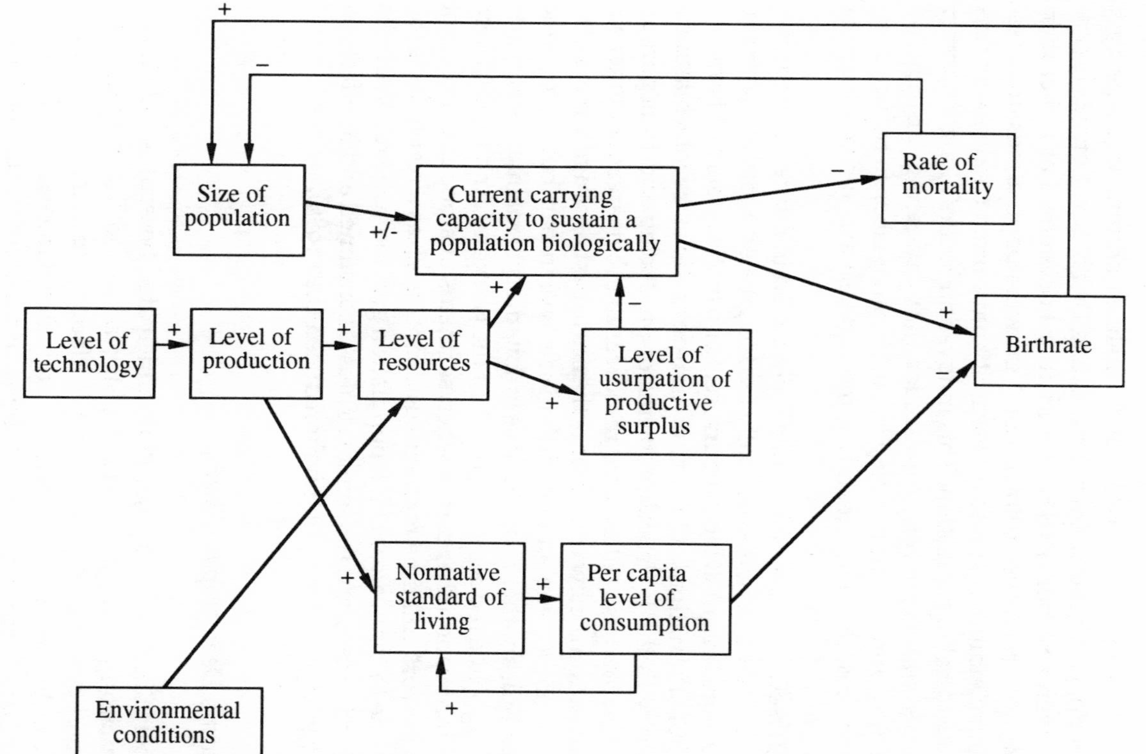

Figure 2.1. Malthus's Model of Population Dynamics

The Boserup Alternative

Esther Boserup (1965, 1981) has formalized arguments made by a number of early sociologists, concluding that innovation has historically been a response to Malthus's four horsemen. For rather than initiating a "Malthusian correction," growing populations under escalating conditions of resource scarcity have often created new patterns of social organization. This innovation occurs along a number of fronts: (*a*) expanding production, (*b*) increasing distribution capacities, and (*c*) proliferating the division of labor to reduce competition and, thereby, expanding the scale, scope, and efficiency of economic activity.

In Boserup's (1965, 1981) model portrayed in figure 2.2, increases in population size will eventually lead to greater population density (i.e., urbanization), which, in turn, sets into motion a series of innovative transformations. One set of transformations revolves around coping with the logistical problems of producing and transporting sufficient food as well as other resources to urban areas. Another set is the creation of new economic activities to sustain wage levels in the face of competition among individuals within particular economic niches. Still another is productive innovations to maintain standards of living while reducing the length of the workday. And finally, there is the innovation that comes when segments of a population do not need to work full time for their sustenance and, as a result, have the free time to innovate. Yet if a population gets too large relative to its productive capacity, economic surplus is consumed and its effects on innovation are correspondingly reduced (Lee 1986).

Spencer's Demographic Theory

Herbert Spencer ([1874–96] 1898) anticipated Boserup's and others' (Carneiro 1967, 1973; Lenski 1966; Maryanski and Turner 1992) analysis of the relationship between population size and social transformation. As is well known, Spencer argued that increases in the size of a population are accompanied by structural differentiation to support and sustain the larger "social mass." What is less recognized, I think, is that Spencer presented a highly sophisticated scenario of

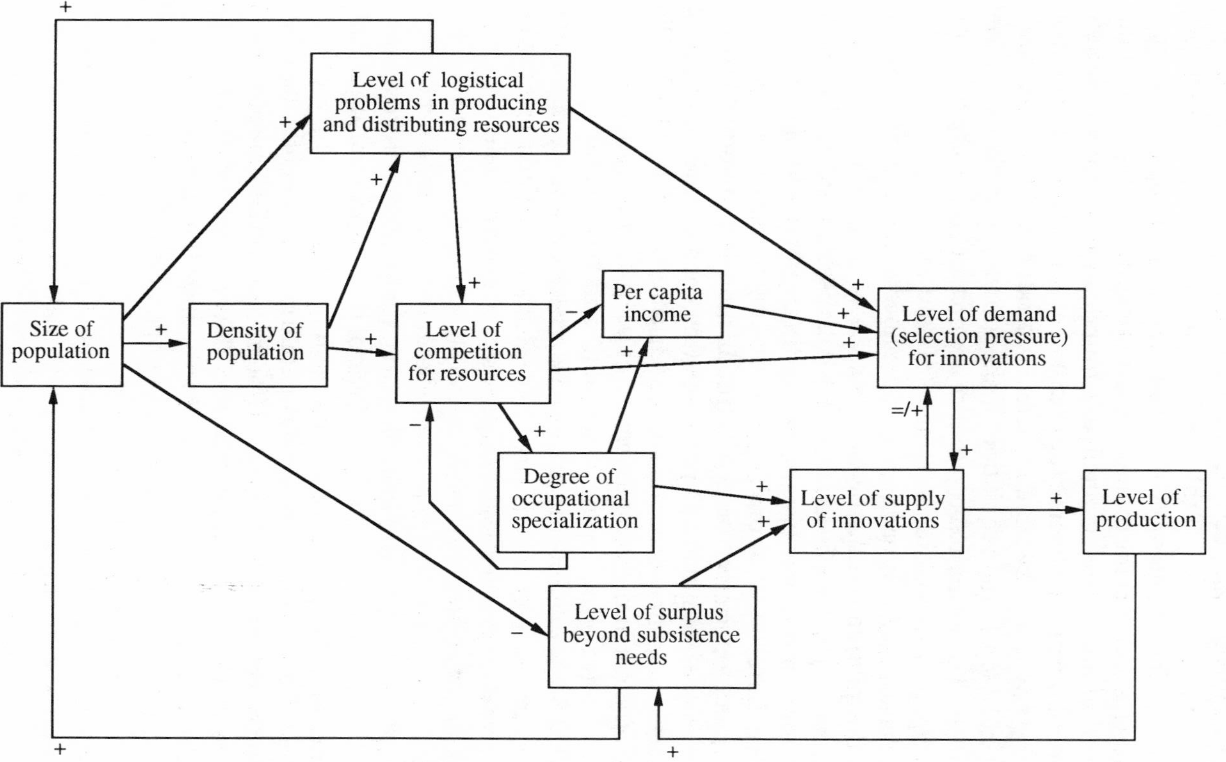

Figure 2.2. Boserup's Model of Population Dynamics

why this relationship between size and differentiation should hold (Turner 1981a, 1984a, 1985a). Spencer's argument is as follows:

As populations grow, two types of logistical loads increase proportionately. One is the problem of maintenance of the growing population, which, in turn, revolves around (*a*) producing sufficient economic surplus to sustain the population and (*b*) reproducing new members who can become sufficiently productive. The other basic logistical load is the regulation of the growing population so that control and coordination can reduce conflict and other pressures for what Spencer ([1862] 1898) termed "dissolution."

If these logistical loads for maintenance and regulation are not met, the population dissolves or suffers at the hands of the four horsemen. Spencer averred that such collapse had occurred often in the course of human history, but at times people found ways to resolve these logistical problems by creating (*a*) new bases for what he termed "operative" processes (production and reproduction), (*b*) new types of "regulatory" processes (the consolidation and use of power), and (*c*) new mechanisms for "distribution" (allowing for the more efficient and rapid movement of people, information, and resources in ways facilitating "operation" and "regulation").

What Spencer implicitly proposed is a selection argument, but one that is very different from Darwinian selection. For Spencer, population growth escalates logistical loads that generate selection pressures *for* (as opposed to *among*) social structures that can meet these loads. If a population cannot create these structures, whether by luck, borrowing, innovation, planning, or some other process, then dissolution will occur. This is the logic of most functional arguments (Turner and Maryanski 1979), and we might term these "functional selection" processes because they operate much differently than Darwinian selection. In recognition of Spencer's contribution, however, I will term these functional processes "Spencerian selection." Such selection occurs in the absence of structures (i.e., low density in a niche), whereas Darwinian selection operates under conditions of high density among units in a niche.

Population growth may generate both types of selection simultaneously. For example, in Boserup's view and, as we will see, Durkheim's as well, urbanization increases the density of resource niches, and thereby sets into motion Darwinian selection via esca-

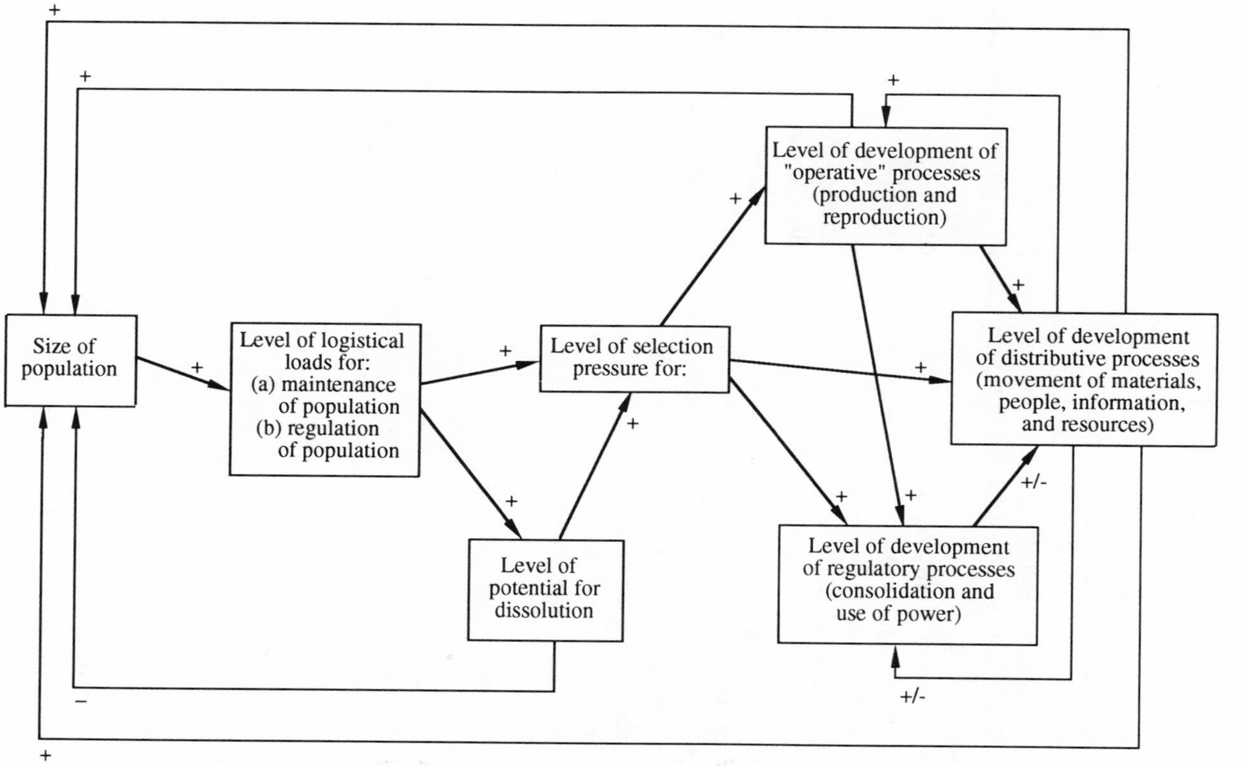

Figure 2.3. Spencer's Model of Population Dynamics

lated competition for resources and consequent specialization of activities (the sociological equivalent of "speciation"); and at the same time, Spencerian or functional selection can operate to generate new productive technologies, new political arrangements, and new mechanisms for distribution, and in so doing, Spencerian selection can generate new resource niches and, hence, increase differentiation. The specter of a Malthusian correction in a population can thus lead to innovations fueled by both Spencerian and Darwinian selection pressures.

As is evident in my rendering of Spencer's argument in figure 2.3, Spencer visualized human populations as constantly facing collapse and dissolution via two scenarios. First, if selection pressures caused by population growth do not result in new operative, regulatory, and distributive structures, then the potential for dissolution increases. Second, even when new structures are elaborated, the positive signs on the reverse causal arrows in the model indicate that further population growth ensues, thereby escalating logistical loads and, hence, the potential for dissolution. Thus, perhaps more than any of the early theorists and even many today, Spencer saw human populations as constantly facing collapse, even as they elaborated and differentiated structures and symbol systems. He never specified the negative reverse causal force—as did later demographic transition theorists—that would check further population growth as new structures and cultural systems were developed in response to growth. Still, despite this shortcoming, Spencer connects conceptually several driving forces—production, power, and distribution—to population growth in a manner that is critical to a theory of macrodynamics.

Durkheim's Ecological Theory

Although Durkheim borrowed a great deal from Spencer (see Turner 1984a and 1985a for documentation), he made explicit the Darwinian metaphor, a metaphor that was pursued by Boserup and more general ecological theories—as will be explored in chapters 6 and 7. For Durkheim ([1893] 1933), population growth increases "material density," especially as transportation and communication technologies reduce "the space" among actors (Turner 1981c, 1990e). Under conditions of density, competition for resources increases and, as a result, the "less fit" seek new resource niches and, thereby, foster specialization or the division of labor. Durkheim thus offers a Darwinian view of the causes of differentiation. This Darwinian view is, of course, only an analogy because selection works on units who can alter their behaviors and transmit such alterations in ways more reminiscent of Lamarckian transmission than genic inheritance. Moreover, Durkheim did not visualize "selection out" or "death" of species; instead, differentiation enables humans to avoid such outcomes as purposeful actors, especially the "less fit," seek resources in other niches or actually create new types of resource niches for themselves. Thus, to compliment Spencerian selection *for* units, we might visualize Durkheimian selection *among* units as a central mechanism of macro-level social organization.

Durkheim's model is somewhat vague because he does not specify, as did Spencer, the types of units and axes of differentiation. But he introduced an important theoretical lead: as population densities escalate, teleological actors are often able to avoid Darwinian "extinction" by constructing new social structures that both utilize and create new resources and, hence, new resource niches for further structural elaboration. Indeed, the prospects of extinction mobilize actors to find or create new resource niches. The organization of populations in space and over time is thus very much an ecological process in which the level and nature of resources, the densities of actors seeking resources, the intensity of competition for resources, and the capacity to generate new resource niches and differentiate new sociocultural systems are all crucial processes. Population growth sets these ecological processes into motion, but they soon take on a life of their own, as will be explored in chapters 6 and 7.

The Dynamics of Population Growth: A Synthesis

What, then, can we take from these approaches that will give us a general model of the effects of the relation between population size and macrodynamics? In figure 2.5, I have outlined in broad strokes my argument. Before discussing this argument, let me repeat what I

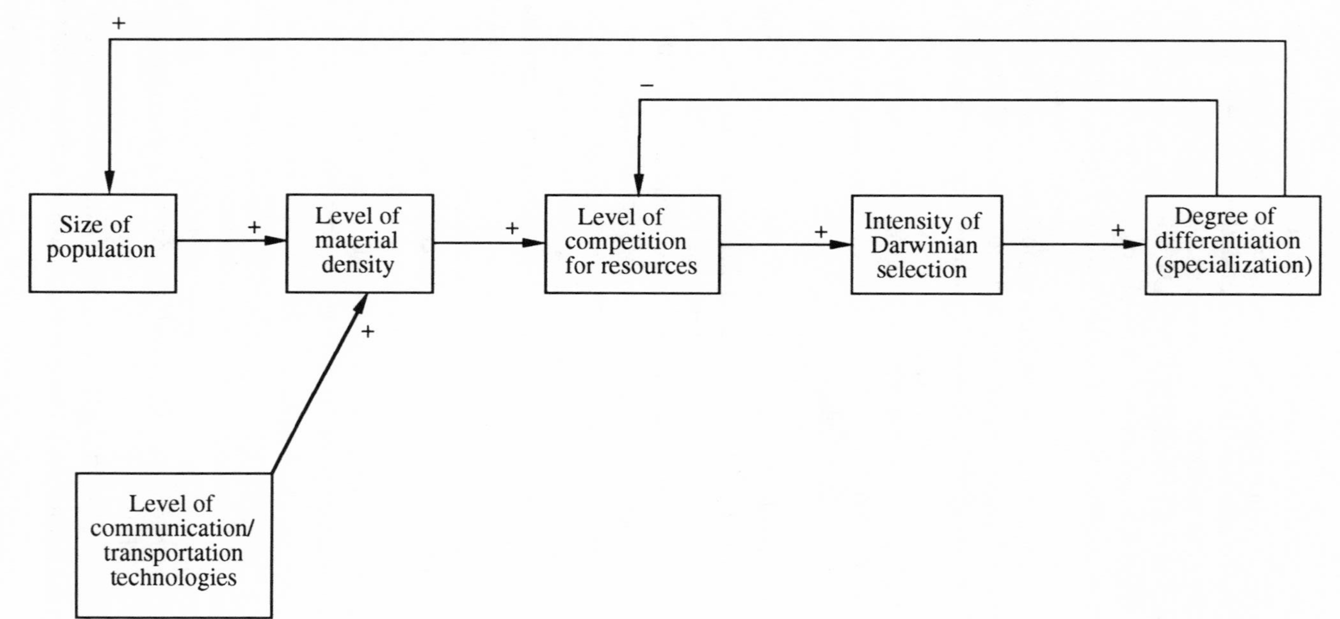

Figure 2.4. Durkheim's Model of Population Dynamics

said in chapter 1: Just because a theoretical model is complex does not mean that it is description dressed in theoretical clothing. All the forces in the model are generic; none refers to a particular population at a specific point in time. Rather, the model delineates a fundamental set of processes operating in all times and places for all populations. Or, at least, this is what I assert.

Population growth causes mounting logistical loads for sustaining a population, while setting into motion potential for a Malthusian correction and/or dissolution. As logistical loads mount, Spencerian selection pressures increase, escalating the potential for dissolution and, at the same time, initiating efforts to (*a*) expand production in order to sustain the growing population, (*b*) consolidate power to regulate, coordinate, and control the population, and (*c*) develop new distributive mechanisms for moving goods, information, and people. If under Spencerian selection pressures (*a*), (*b*), and (*c*) can occur, then they operate through the reverse causal paths delineated in the model to reduce what I am terming "first-order" logistical loads, or those loads for sustenance, maintenance, and control that can cause directly a Malthusian correction or, via their effects on Spencerian selection, the rapid dissolution of the population. As this potential for dissolution is lowered through the increases in (*a*), (*b*), and (*c*), a major obstacle to further population growth is removed. In particular, increased production, as it effects and is affected by power and distribution, generates the material surplus to sustain a larger population, thereby lowering the potential for a Malthusian correction.

As production and distribution expand and increase the level of material surplus, new systems of cultural symbols eventually begin to compensate for the removal of structural obstacles to population growth. For expanded production and distribution, as these increase material wealth, escalate normative standards of living which, in turn, create incentives for lower birthrates—as Malthus hinted and as "demographic transition" models predict. Such is particularly likely to be the case if the consolidation of power creates stable and predictable systems of social services and redistribution which provide income security for those who might otherwise have more offspring as a source of financial security. These systems are most likely with high production and distribution which create the resource

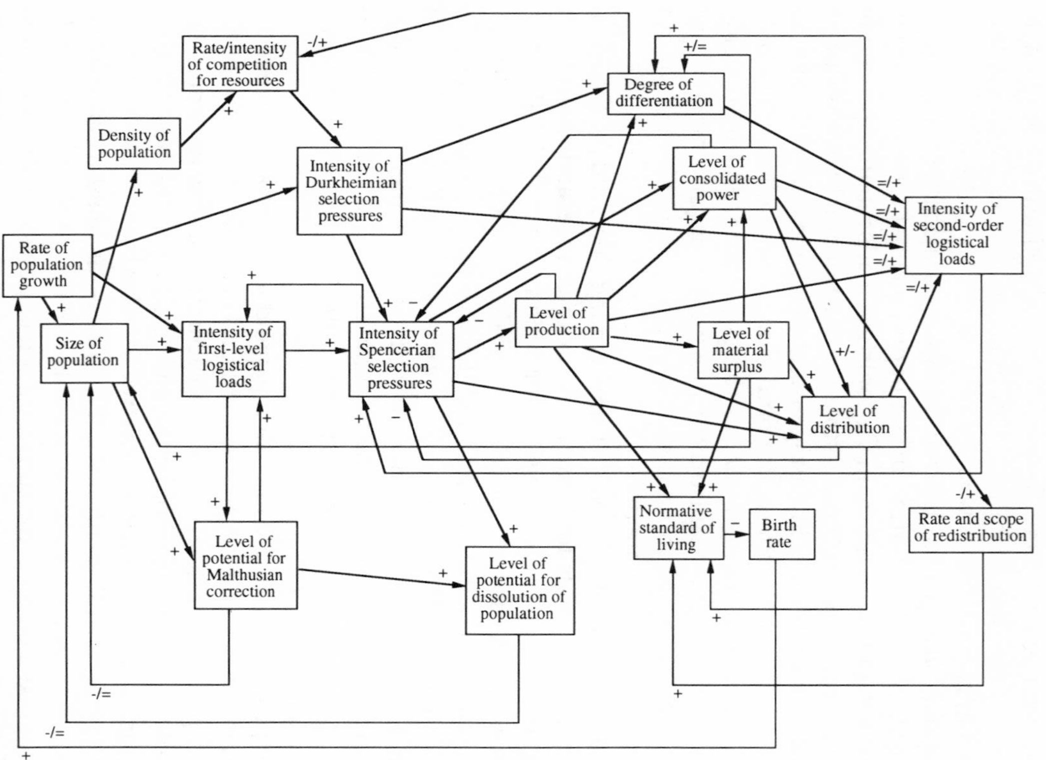

Figure 2.5. Population Size and Macrodynamic Processes

surplus as well as the democratization of political processes responsive to population demands. When centers of consolidated power cannot, or will not, redistribute and provide such security systems (as was the case in most of world history up to the nineteenth century and as is the case in much of the Third and Fourth World today), population growth will continue up to the point of a Malthusian correction and/or population dissolution as each member uses procreation of offspring as a source of future security. These events have been subject historically to contingent events, such as wars, plagues, crop failures, and the like. Yet even these events are somewhat predictable: wars increase consolidated and centralized power, especially under conditions to be examined in chapter 5; disease and plagues are, of course, one of Malthus's predictions, but even when population pressures on resources have not reached the point of exhaustion, these events can occur under conditions of high-density urbanization in low-technology populations; and while crop failures can occur in any type of society, their effects on population processes are less in high-technology systems that have access, via markets or conquest, to the resources of other societies.

In addition to first-order logistical loads, the model in figure 2.5 denotes what might be termed "second-order" logistical loads, which are caused by the elaboration and differentiation of productive, distributive, and power structures. These structures may be created, at least initially, in response to first-order logistical loads stemming from population pressures. For, although expanded levels of production, distribution, and power-use (see chapters 3, 4, and 5, respectively) lower the level of first-order logistical loads caused by population growth, they generate new kinds of logistical loads that are less directly caused by population processes. Thus, as production becomes ever more complex, as distribution becomes more voluminous, extensive, and rapid, and as power becomes more consolidated, centralized, and extensive in its regulatory activities, new kinds of logistical problems begin to emerge. These problems tend to revolve around coordination, conflict, and control of units within and between the productive, distributive, and political sectors of a society as well as the additional sociocultural units (such as social classes, ethic subpopulations, political parties, rival interest groups and organizations, social movements, conflict groups and organizations) created by the elaboration of production, distribution, and power. And so, as the specter of a Malthusian correction and imminent dissolution are reduced, new logistical problems emerge.

If the integrative problems associated with second-order logistical loads become intense, they dramatically escalate Spencerian selection pressures, thereby increasing the potential for dissolution as a result of a collapse of key productive, distributive, and power structures. For example, currency can become so inflated as to cause a collapse of distribution and a return to local barter markets; centers of power can become so corrupt, so ridden with fiscal crises, or so unresponsive to inequalities that revolt and revolution break a population apart; centers of power can also be so inept in geopolitical relations that conquest and destruction by other populations ensue; or productive units can exhaust resources, engage in cutthroat competition, and encourage polarization of class and ethnic antagonisms to the point that they cease to be reliable sources of population maintenance.

All of these disintegrative tendencies associated with second-order logistical loads are several steps removed from those first-order loads caused by population size and growth. And yet, they are never far removed. In fact, as Durkheim recognized, dense settlements and competitive social relations are not only a frequent battleground for conflicts stemming from second-order logistical loads but also the cause of these second-order problems of integration. For as density and conflict over resources cause sociocultural differentiation (see chapter 7), problems of integrating a more diverse and divided population increase. And because population size, growth, and movement set into motion centrifugal forces revolving around distribution, power, and production that, in turn, feed back to affect population processes, population size is, once again, not many steps removed from second-order logistical loads associated with differentiation.

Thus, both first- and second-order logistical loads increase Spencerian selection pressures, with the result that populations always face potential dissolution *unless* they find new ways to: (*a*) organize production, distribution, and power and (*b*) manage the problems of coordination, control, and conflict that accompany differentiation of activities along these three axes. One consequence of increased levels of production, distribution, and power is, as Malthus hinted and demographic transition theorists have emphasized, a rise in normative standards of living, which, in turn, tend to lower birthrates. If this causal connection is supplemented by redistribution of some productive surplus and implementation of security benefits by centers of power to the mass of a population, then an additional force operates to reduce birthrates and, as a consequence, decrease population size or at least its rate of growth (as is denoted by the causal paths from normative standard to birthrate and, then, to rate of population growth). In turn, first-order logistical loads are reduced, thereby dampening those Spencerian and Durkheimian selection pressures emanating from population size and growth. Still, new second-order logistical problems emerge as a result of new productive, distributive, and political arrangements, but these are less directly connected to population as a driving force and to the potential for a Malthusian correction. Dissolution is still possible as a result of unresolved second-order problems, but this potential is less driven by population than by the subforces, such as inequality, ethnic diversity, geopolitics, and other derivative forces that ensue from increased production, distribution, and power-use (as will be explored in chapters 3, 4, and 5).

Population Diversity and Macrodynamics

I have emphasized the importance of population size because, in my view, it is *the* most important demographic dynamic. Yet I should at least mention other population processes, each of which intensifies and accelerates the dynamics outlined in figure 2.5. One of these processes is population diversity. In general terms, as populations become more differentiated, second-order logistical loads escalate, especially those revolving around regulation, coordination, and control, because of the potential conflicts among diverse subpopulations. And if diversity reaches very high values and cannot be regulated, dissolution of the population becomes more likely.

Diversity is systematically generated by production, distribution, and power-use. As production and distribution increase, so does the division of labor creating distinctive subpopulations of individuals. Production and distribution also generate inequalities in the amount and nature of resources received by individuals, thereby creating new bases for differentiating subpopulations in terms of their respective shares of resources. The consolidation and use of power also cause differentiation in terms of where subpopulations stand in relation to centers of power and in terms of expanding the division of labor (with respect to the administration of power). The consolidation of power also aggravates inequalities, as those with power usurp more resources, and as a result, new kinds of divisions emerge in society. Power is frequently used in geopolitical conquest and annexation, which further increase not only the size but also the diversity of a population.

The end result of these forces generating diversity in the characteristics of a population is to create new logistical burdens that dramatically increase the potential for dissolution and, at the same time, escalate Spencerian selection pressures which, ironically, can lead to patterns of production, distribution, and power-use that increase diversity and potential for dissolution.

Population Distribution and Macrodynamics

The geopolitical distribution and migration patterns of a population will also have implications for macrodynamic processes. As indicated in the model in figure 2.5, the density of a population increases both Durkheimian and Spencerian selection pressures, but more is involved. The size of territory, the number and density of settlements, the distances between them, and the correlation of territory or settlements with culturally and organizationally diverse subpopulations have implications for macro-level processes. As will be outlined in more detail in chapter 6, these forces increase logistical loads, especially for power-use and distribution. And these loads are dramatically escalated when they are associated with inequalities and past geopolitical conflicts.

Migration patterns also influence macrodynamic processes. Immigration increases the size of a population, and because most immigrants move into dense settlements, immigration increases density of the population as well. Thus, immigration increases both Spencerian and Durkheimian selection pressures; and if immigration is associated with inequalities or geopolitical activities, its effect on logistical loads is increased. Emigration decreases population size if it exceeds birthrates and immigration, thereby potentially lowering Durkheimian and Spencerian selection pressures. Yet if emigrants are members of the more productive sectors of a population, then immigration can increase logistical loads, even for a smaller population.

Internal migrations usually increase logistical loads through their effects on population density and on reorganizing productive, distributive, and regulatory processes to accommodate the redistributed population. Most internal migrations are to urban regions, thereby increasing population densities and escalating selection and its effects on both first- and second-order logistical loads to generate new productive niches, distributive facilities, and control structures that can accommodate and manage the new migrants. When such migrations cannot be absorbed into productive structures, as is the case with the larger "shanty towns" in much of the undeveloped world today, then logistical loads are dramatically escalated, especially as they are increasingly fueled by inequalities. Even without its effects on population density, large-scale internal movements increase logistical loads by virtue of the need to redistribute productive capacities, to provide distributive infrastructures, and to reorder power-use to manage the realities of a new demography for a population.

Elementary Principles of Population Dynamics

Population processes work through other basic forces, as is indicated in figure 2.5. Thus, by itself the analysis of population dynamics may seem strained. Yet it is essential to highlight population as a force behind those other social processes—e.g., power, production, inequality and stratification, differentiation, and conflict—more typically emphasized by sociologists. With this caveat, the basic line of argument presented in this chapter can be summarized in a series of elementary principles:

I. The size and rate of growth of a population is:
  A. A positively curvilinear function of the level of material surplus available to support members of a population, which, in turn, is a positive and multiplicative function of:
    1. the level of production
    2. the level of distribution
    3. the rate of redistribution from centers of consolidated power
  B. A lagged negative function of the modal normative standard of living among a population's members, which, in turn, is a positive function of conditions A-1, A-2, and A-3

II. The potential for dissolution of a population is a positive function of Spencerian selection pressures, which, in turn, are a positive and additive function of:
  A. First-order logistical loads, which are a positive and additive function of:
    1. the absolute size of a population
    2. the rate of population growth
    3. the potential for Malthusian correction, which is a positive function of a population's size relative to its productive capacity
  B. A lagged positive function of second-order logistical loads, which are a positive and multiplicative function of:
    1. the level of consolidated power
    2. the level of production
    3. the level of distribution
    4. the level of differentiation
    5. the level of Durkheimian selection
  C. A negative function of the level of production, consolidated power, and distribution up to the point where second-order logistical loads pass a threshold point and activate conditions listed in II-B above
  D. A positive function of the intensity of Durkheimian selection pressures, which, in turn, is a positive function of the level of competition among members, which is a:
    1. lagged positive function of population density

2. negatively curvilinear function of the level of differentiation

III. The level of production and consolidated power evident for a population is a positive function of the intensity of Spencerian selection pressures, which, in turn, is a positive function of:

A. First-order selection pressures, which are a positive function of the conditions listed in II-A

B. Second-order selection pressures, which are a positive function of the conditions listed in II-B

3

PRODUCTION DYNAMICS

LIKE ALL LIFE FORMS, humans must extract resources from the environment, convert them into usable commodities for sustaining life, and then distribute them to members of the population. Those populations in human history that could not effectively organize these basic processes were selected out, whereas those which could were able to survive and reproduce themselves. Reproduction of a population is thus dependent upon production dynamics, or the gathering of resources, the conversion of resources into usable goods, and the distribution of goods (Turner 1972).

In approaching the dynamics of production, I will stress in this chapter gathering and conversion processes, saving for the next chapter a more extensive analysis of distribution dynamics. For as I will argue, distribution becomes a driving force in its own right, somewhat independently of gathering and conversion; and in fact, when distribution processes are highly developed, they become a more significant force than production, influencing not only the amounts of resources gathered and converted but also other basic macro processes, such as the level of power, the size of a population, the distribution of a population in space, the differentiation of a population, and the disintegrative and integrative forces evident in the organization of a population.

The Nature of Production

Basic Elements

Production involves bringing to bear on gathering and conversion a number of basic elements (Turner 1972:20–23): (1) technology, or knowledge about how to control and manipulate the environment, both natural and social; (2) physical capital, or implements that are used to gather, convert, and distribute as well as the liquid resources used to acquire such implements; (3) human capital, or the number and characteristics (knowledge, skills, motivations, etc.) of the individuals; (4) property, or the socially constructed right to own, possess, and use objects of value; and (5) entrepreneurship, or the mechanisms available in a society for organizing technology, physical capital, property, and human capital in gathering, converting, and distributing (Parsons and Smelser 1956). Figure 3.1 delineates the relations among these basic elements.

The Level of Technology

As knowledge about how to manipulate and control the environment, both physical and social, technology directly affects the levels of human and physical capital as well as the kinds of organizational or entrepreneurial mechanisms available to coordinate technology, capital, and property in gathering, converting, and distributing. As is evident in the model, there are important reverse causal effects from these forces affected by technology. The level of human capital will determine the level of effort in generating knowledge and innovation that will be available for production; the extent of physical capital will determine the resources—both material and liquid—that are available for creating and then implementing new technologies; the nature of the available organizational mechanism for coordinating productive and distributive activity will influence the rates of technological innovation as well as the way in which they are brought to bear on production; changes in nature, amount, and diversity of property and associated social relations will stimulate innovation

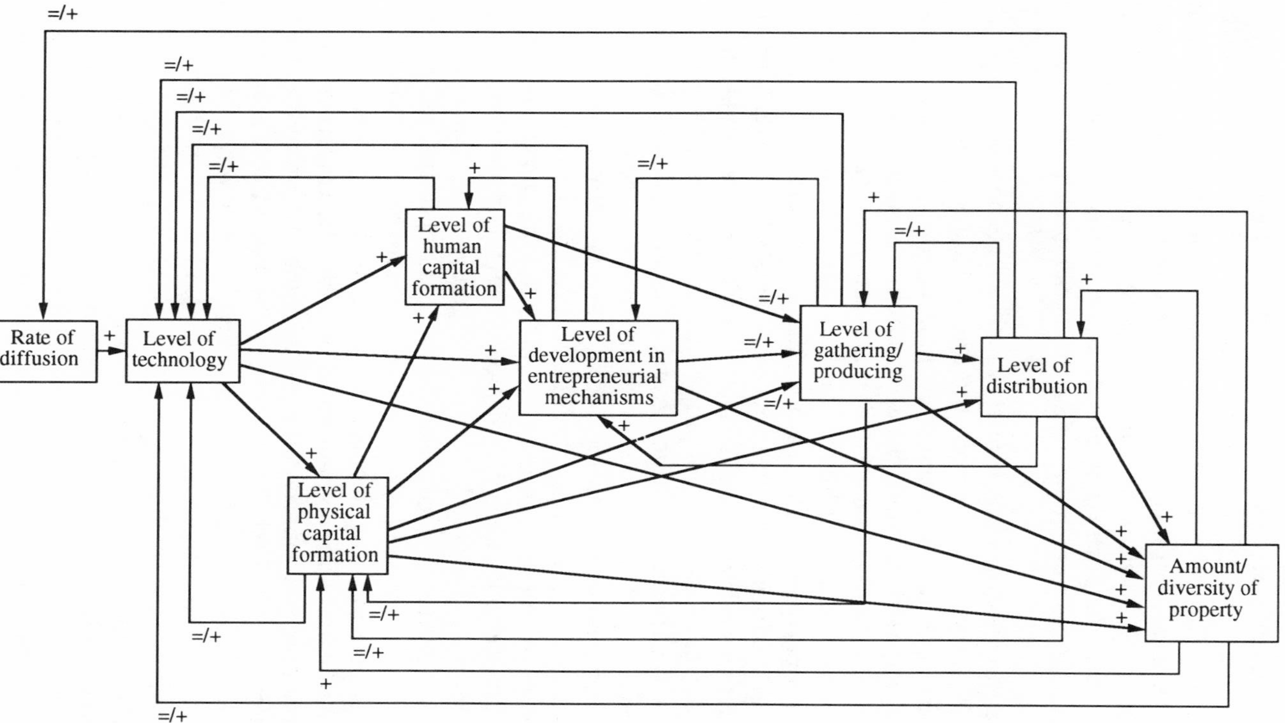

Figure 3.1. Dynamic Elements in Production

and its diffusion across productive sectors; and the actual process of gathering, producing, and distributing often leads to new technological ideas, although these productive elements can constrain and limit the development and use of new technologies until they reach very high levels. Aside from these direct effects back to technology, the model portrays numerous indirect effects as these forces influence one another in working their way back to the level of technology.

These reverse direct and indirect causal effects on technology are almost all lagged because until the forces delineated in figure 3.1 reach very high levels, populations tend to get locked into "technological regimes" in which the application of a given level of technology leads to the use of a limited range of entrepreneurial mechanisms to organize human capital, physical capital, and systems of property that sustain a population in a particular mode of gathering, producing, and distributing. This is why the signs for the causal arrows are lagged positives (=/+); it takes a certain threshold of capital formation, diversity of property relations, and entrepreneurship to raise gathering, producing, and distributing to new and dramatic levels; and it is only with industrial capitalism, where competition and concerns for profit systematically generate the search for new ways of doing things, that technological innovation becomes built into the very nature of gathering, producing, and distributing (Lenski, Lenski, and Nolan 1991). For most of human history the rate of technological innovation was very slow, but as technology gradually accumulated as a result of diffusion, chance, and innovation (usually stimulated by new problematic conditions, such as population growth or ecological change), new levels of capital formation and entrepreneurship were reached, as were new and ever more diverse property relations, with the result that the rate of innovation began to accelerate. Thus, hunting and gathering technological "regimes" were locked into low levels of physical capital (spears, bow and arrow), relatively low skill levels in the human capital (knowledge of hunting-gathering), simple entrepreneurial mechanisms (nuclear families in bands), and few forms of property (because all property and capital could be created by individuals themselves, all property had to be transported by hand, and all land was collectively "owned") with the consequence that gathering, producing, and distributing remained at low levels (Service 1962a, 1962b, 1966; Lee 1979; Lee and DeVore 1968, 1976; Elkin 1954; Sahlins 1972). There was no incentive to change, nor internal tensions forcing change; and hence, until population growth among more settled hunter-gatherers generated Spencerian selection pressures for the application of new technologies, new kinds of physical capital, new human capital requirements, new entrepreneurial mechanisms, and new types of property, the nature of production remained much the same for thirty thousand years (Maryanski and Turner 1992). With horticulture, there were more pressures for change because the entrepreneurial forms (revolving unilineal descent linking kin units) were inherently tension provoking due to the forced proximity of individuals in large kin networks (Maryanski and Turner 1992). Similarly, advanced horticulture and later agrarianism generated tension-producing entrepreneurial structures (centralized political authority, coupled with higher levels of inequality over property) that would, eventually, generate Spencerian selection for new entrepreneurial forms, new forms of capital, new types of property relations, and, ultimately, new levels of technology (Bender 1975; Curwen and Hatt 1961; Flannery 1973; Bloch 1962). Hence, the general point is that technological innovation was for most of human history slowly accelerating with periodic jumps to new "regimes" whose structures then sustained a population in a particular productive mode until the inherent tensions in entrepreneurial mechanisms, in the distributions of power and property, or in other forces such as population growth generated selection for new technological innovations.

Physical Capital

The implements of production, as emphasized above, are reflective of technology, but more is involved. In the course of using capital, new innovations occur; and although entrepreneurial processes as well as other features of the broader culture and social structure may limit (or accelerate) their adoption and use, members of populations always develop new implements. Such processes are dramatically escalated by contact, of whatever kind, with other populations whose technologies and tools will eventually be incorporated into production (Hawley 1986). Thus, the physical capital of a population gradually increases; and even if war, conquest, and plunder temporarily destroy capital, diffusion of the technologies that guide capital formation will in the long run increase the level of capital unless, of course, a population is completely destroyed.

Distribution processes also work to increase physical capital. One line of influence is through the expansion of trade with other populations, which, in turn, increases the flow of technologies and implements. Another is through the volume and velocity of markets, to be examined in more detail in the next chapter, which increase not only those gathering and conversion processes requiring new forms of physical capital but also those processes generating and extending the use of liquid capital—money and all the many financial instruments that money makes possible (Collins 1990; Earle and Ericson 1977; Braudel [1979] 1982; Mann 1986). Moreover, high volume and velocity markets transform property relations, making them increasingly transferable and transformable in response to market forces. Yet another influence is the creation of the physical infrastructure for transportation, communication, and trade in markets, which, in turn, extend the use of liquid capital (some of which will be used to build additional physical infrastructure).

Human Capital

Like physical capital, the number of people available for productive activity, as well as their characteristics, reflect the nature of technologies as these create those physical implements and entrepreneurial structures used in production. Population growth creates Spencerian selection for expanded production, and with increased production, a larger population can be sustained. Thus, in a gross calibration, the size of the human capital pool and the level of production are reciprocally connected, as I stressed in the last chapter.

The characteristics of this pool—that is, its skills, motivations, orientations, and the like—reflect the way physical capital is organized by entrepreneurial mechanisms. If physical capital is organized in simple forms, such as pooled networks of kin working with simple tools to harvest gardens (horticulture), then the diversity of skill will be relatively low, and motivations, orientations, and other attributes will be focused on a narrow range of tasks done in traditional ways. If, on the other hand, physical capital is organized in factories and bureaucracies linked together by diverse markets, then the characteristics of the human capital pool will be diverse, as will their motivations, orientations, and other characteristics. In turn, a more diverse pool of capital creates (*a*) expanded ranges of niches for new productive goods and services to meet varying tastes and needs, and (*b*) innovative possibilities for new technologies emerging out of the synergy of differentiated labor working in complexes of physical capital organized by a wide range of entrepreneurial mechanisms.

Entrepreneurial Mechanisms

The notion of entrepreneurship has a somewhat narrow meaning in traditional economics, but from a sociological perspective, the concept should be broadened considerably to include *any* significant feature in the culture and structure of a population that affects the organization of technology, physical capital, human capital, and property in gathering, producing, or distributing (Parsons and Smelser 1956). Thus, for example, corporate bodies, cartels, tort laws, regulatory agencies, free markets, courts, kinship norms, monetary systems, credit and insurance systems, banking systems, roadways, communication networks, and many other features of a social system are entrepreneurial *to the degree* that they organize technology, capital, and property in gathering, producing, and distributing. This shift to a broader conception is, I feel, necessary when analysis moves beyond the internal workings of corporate actors in market systems to the sociocultural forces organizing whole populations in many different types of social systems.

Entrepreneurial processes operate at several different levels to organize: (1) the structural units in which gathering, conversion, and distribution occur, (2) the relations among such productive and distributive units, and (3) the relations between productive units, on the one side, and individuals and other nonproductive structural units, on the other side. Let me examine each of these separately.

The nature of the structural units within which productive activity is organized reflects the level of technology, capital formation, and nature of property relations, but these structural units also place limits on technological development, capital accumulation, and property relations. If a band, a kin group, or a feudal manor are the principal structural units, as well as the arbitrators of property relations, then the capacity to develop new technologies and new forms of capital will be highly constrained. If, however, a profit-making corporation is the major type of structural unit organizing activity, then fewer constraints on technology and capital, both human and physical, exist. And, property relations change with the ebb and flow of profits and use of capital in market systems. Thus, as noted earlier, part of the reason that populations get locked into technological regimes is the conservatism and constraints of the units within which production occurs, especially the definitions over property where some actors secure control of objects and symbols used in production and, as a consequence, seek to sustain the privilege that comes with such control.

Another part of the reason is the ways in which relations among productive units are organized. If productive units consume rather than exchange their own productive outputs, this fact places limits on the types of relations that can exist among units. Such consumptive productive units do not need as many mechanisms for regularizing relations with other units as do those generating outputs for exchange. Hence, tradition and external political authority can usually manage relations; and because these are conservative forces, they dampen technological innovation. In contrast, once exchange of commodities and eventually services among structural units exists, new entrepreneurial mechanisms for regulating the exchanges emerge as a result of Spencerian selection. Tort law, courts, administrative agencies attached to political authority, symbolic media like money, and markets all evolve to regulate more complex relations and, in so doing, create opportunities for new types of structural units, new forms of physical and human capital, new forms of property relations, and new technologies. And once such an expanded array of entrepreneurial mechanisms mediating relations among productive units is in place, it operates to accelerate all aspects of production, and as we will see in later chapters, other facets of a society.

The elaboration of these kinds of entrepreneurial mechanisms expands the range and diversity of relations not only among productive units but also among all types of units organizing the members of a population. The use of formal laws, adjudicative courts, administrative agencies, money, market exchanges, and moderated enforcement capacities of centralized political authority all work to create a wide range of fluid and flexible connections among individuals and structural units. In so doing, they increase the number of niches and opportunities for innovations and for elaboration of social structures within and outside the sphere of production to ever more complex forms (see chapter 7).

Property

The notion of "property" is somewhat elusive, but nonetheless fundamental. For my purposes, I will define property as the right to possess and, by virtue of such possession, to control and use objects of value. All elements of an economy—technology, capital, entrepreneurship—are or can become property, as can any resource or object that is valued by the members of the population. The embeddedness of property in so many facets of social life is what makes it elusive; it is potentially everywhere but hard to isolate from those processes in which it is instantiated.

What makes property a crucial feature of human populations is that, as the level of production and distribution rises, the amount, diversity, and movement of property increase. As this occurs, a higher proportion of relationships and transactions involve property, or the right to possess, control, and use objects of value. And once property relations become a central feature of an economy, as well as the society at large, they accelerate the values for other economic processes delineated in figure 3.1. Physical capital formation, for example, is increasingly a matter of property rights over the implements of production and the social relations of production; human capital is increasingly defined in terms of rights to labor-power for specified periods of time (with slavery being an extreme case of possessing the entire human being as property); entrepreneurship is increasingly concerned with securing and coordinating the property rights for technologies, physical and human capital, and resources necessary for production. Thus, patents for technologies, contracts for labor, ownership of equipment and land are all manifestations of property rights; and the nature of such rights determines how gathering, producing, and distributing will occur. Property thus becomes a kind of common denominator of production. It provides a basis for defining control over objects, symbols, and social relations as well as for keeping account of who has what and, equally significantly, who does not have what.

As is emphasized in figure 3.1, the absolute amount of property (i.e., the proportion of all objects of value that individuals and collective units can possess) and the diversity of property (i.e., the number of different types of objects that are defined as property) are viewed as its crucial macro-level features. For as the amount and diversity of property increase as a causal effect of the other elements listed in the model, property becomes increasingly a force in its own right, defining objects of value which can create demand in markets for new goods and, in turn, feeding back on gathering-producing, entrepreneurship, capital formation, and technological development. And most importantly, the capacity to define property provides a crucial link between production and the dynamics of power, for property is embedded not only in power and privilege but also in deprivation, tension, and conflict.

For much of human history, property was not a way of defining the world; and even as property began to emerge as a construct for defining objects, property was collectively held or, if given to an individual (such as a headman in a sedentary gathering population, or a kinship leader in a horticultural system), obligations for redistribution mitigated against views of objects as property, especially "private" property. But once a certain threshold is reached in the proportion of objects that can be *possessed* by actors, property becomes a self-escalating force. Those who do not have property want some, creating demands for innovations and new productive outputs. In turn, these new outputs of objects begin to diversify the nature of property. And eventually, as is the case in modern capitalist systems, property rights penetrate all relations and virtually all objects are defined in terms of property considerations. This "commodification" of the social life is often used to mount critiques of capitalist systems, but in fact, commodification is the result of a long historical trend toward increasing the amount and diversity of objects seen as property. Market systems have simply accelerated this trend, especially open markets employing generalized media (see chapter 4).

Production and Power

Power and production are intimately connected: the consolidation, centralization, and administration of power are not possible without sufficient productive surplus to support those structures involved in power-use; and conversely, the formation of physical capital (e.g., productive and distributive infrastructure) and the development of entrepreneurial mechanisms (e.g., money, law, courts, and markets) are dependent upon the regulatory capacities of centralized political authority. And most significantly, once an economic surplus exists, property as a social dynamic becomes ever more salient, because its "ownership" must be designated, and as Marx ([1848] 1978, [1867] 1967) emphasized in more strident terms, power increasingly revolves around the distribution of property and the social relations that inhere in possession of property.

Increasing production not only generates the surplus necessary to support centers of power (Lenski 1966), it does more: expanded production creates new logistical loads—both directly and indirectly through its effects on distribution—for coordination and control of the new property systems created by the existence of a surplus, the new organizational forms made possible by surplus and property, the larger populations that ensue with escalated production, the increased volume of distribution, and the many other forces that are unleashed when production expands. These mounting logistical loads set into motion Spencerian selection processes for increased regulation. The result is increased concentrations of power, the effects of which are, for the most part, positively curvilinear for production and distribution. Initially, the consolidation, centralization, and use of power increase the rate of technological innovation, the level of development in physical capital, the diversity of human capital, the amount and diversity of property, the viability of entrepreneurial mechanisms, and the level of production and distribution. For centers of concentrated power have an interest in some degree of (*a*) technological innovation, if only for military technologies, (*b*) new productive activities that can generate the material surplus to support the centers of power, (*c*) new objects of possession (i.e., property) that can mark power and, as is often the case, be used to determine "winners" and "losers" in status-prestige competition among elites, (*d*) new entrepreneurial tools to coordinate and control activities so that a stable order and constant flow of material surplus to centers of power can be sustained, (*e*) expanded distributive systems for new sources of surplus that can be usurped and used for expanding the scope and extent of political control, and (*f*) new infrastructural facilities for both housing and moving governmental functionaries.

Yet as power becomes ever more concentrated, a reversal of its effects on these processes begins. This reversal is an outcome of the growing inequality in the distribution of property that inevitably accompanies high levels of concentrated power and the resulting need to sustain the privilege of elites and control the restiveness of non-elites (Lenski, Lenski, and Nolan 1991; Maryanski and Turner 1992). These processes become most evident in agrarian systems in which a feudal monarchy and nobility get trapped in cycles of usurpation of ever more surplus to finance privilege for elites, control of property and the corresponding social relations, external warfare, and internal control of increasingly restive masses (Goldstone 1990; Skocpol 1979). As this escalating process ensues, the growing concentration of power and the increasing inequality in the distribution of property and other valued resources decrease incentives for technological innovation, capital formation, increased production, and escalated market activity because, in the end, increasing proportions of any additional surplus will be usurped by elites and centers of power as *their* property. Moreover, as centers of power exercise ever greater control of potential dissent resulting from the extraction of surplus and the increasing levels of inequality over property, entrepreneurial mechanisms—law, courts, monetary systems, administrative agencies—are used to usurp surplus and control potential conflict, thereby reducing the positive effects of these mechanisms in stimulating increased production and distribution.

Thus, as will be explored in chapter 5, power bears a contradictory relationship to production and distribution: on the one hand, it can facilitate control and coordination of productive and distributive processes in the face of mounting logistical loads while operating as a force behind technological innovation, capital development, and entrepreneurial activity, whereas on the other hand, power tends to become ever more concentrated and to foster tension-producing inequalities and property relations that lead to the use of power in ways that reverse or, at the very least, level out these positive effects. Maintaining this balance between over- and underregulation is always difficult, if not impossible, in the long run (Block 1980).

Production and Population

The size of a population, along with its characteristics and movements, are important elements in production and power dynamics. One line of influence from population growth is its effects on increasing logistical loads that, in turn, generate Spencerian selection for increased production and concentrations of power in order to avoid a Malthusian correction or dissolution. For if sufficient resources cannot be gathered and converted to support the larger population, Malthus's four horsemen will ride again; or if the escalating problems of controlling and coordinating this population, especially as resource scarcity increases, cannot be met, Spencer's ([1862] 1898, [1874–96] 1898) dissolution scenario is a likely outcome. Thus, as Boserup (1965) emphasized, population growth can become a stimulus to innovation that can cause a shift in "technological regimes" and levels of production, or as Spencer argued ([1874–96] 1898), it can be a stimulus for political consolidation that can, at least initially, facilitate the development of new modes of social organization in general and productive organization in particular.

A second general line of influence from population growth is more indirect, operating via its effects on the relationship between power and production dynamics. One set of such indirect effects is this: As population growth causes the consolidation of power, the use of such power initially facilitates the development of productive processes, but population growth also initiates the consolidation and centralization of power that will increasingly extract productive surplus, discourage technological innovation, increase tensions associated with exploitive regulation and inequality over property, and thereby cause not only a "breakdown" of political authority but also a disruption of productive activity. Yet another set of indirect effects is: As population growth causes the expansion of production, the availability of surplus facilitates the expansion of consolidated power and the dynamics of power-use noted above, but it also sustains a "Malthusian pressure" with the ever-present threat of overpopulation and political turmoil. For if production does not keep pace with population growth, political turmoil will ensue; and this potential, as it is manifested in sporadic collective outbursts, accelerates somewhat contradictory activities by political authority. For on the one hand, the development of new entrepreneurial mechanisms, new technologies, new property relations, and new forms of capital expands production and sustains the population, whereas on the other hand, new means of social control and regulation of a restive population have the opposite affect on technological, entrepreneurial, and capital growth.

A third line of influence is the effects of population growth on the consumption of productive surplus, which, in turn, affects the dynamics of production and power discussed above. A growing population consumes more, and as a result, decreases the available surplus which can be used to sustain political authority and productive activity. With respect to political authority, what a population consumes is unavailable for consolidating political power, thereby draining centers of power of needed revenues to maintain control and foster productive expansion and, as a consequence, hastening the potential of political upheaval and dissolution (Goldstone 1990). In regard to productive activity, the surplus that a population consumes is no longer available for supporting technological innovation, for expanding the physical capital base, and for entrepreneurial activity which decreases the capacity of production to keep pace with population growth, thereby hastening a Malthusian correction and/or a Spencerian dissolution.

Thus, population dynamics generate pressures on production directly, and indirectly, through their effects on the consolidation and use of power as it interacts with productive processes. As I noted above, we might call this a "Malthusian pressure" because it presents the specter of insufficient food (the "famine, pestilence, and disease" side of Malthus's argument) and inadequate political regulation (the "war" side of Malthus, if we include internal conflicts within a population as war).

An Elementary Principle of Production Dynamics

Without going into detail over the causal processes—direct, indirect, and reverse—discussed in this chapter, it is possible to summarize, in broad contours, my argument in one proposition:

I. The level of production evident for a population is a multiplicative function of:

A. the positive effects of the level of technology

B. the positive effects of the level of capital formation

C. the positive effects of the level of development of entrepreneurial mechanisms

D. the positive effects of increased amounts and diversity of property and the social relations inhering in the control and use of property

E. the positively curvilinear effects of the level of consolidation and use of power

The multiplicative nature of the relations specified in this proposition glosses over the complex sets of causal processes examined in more detail earlier, but as a rough approximation of the basic dynamics of production, the proposition captures the essential point. If we wanted to discuss gathering, producing, and distributing separately, the same relations would hold, although we would add that each of these processes is positively and multiplicatively related. That is, an increase in either gathering, producing, or distributing accelerates the levels of the other two; for each has effects on technology, capital formation, entrepreneurship, formation of property, population size, and concentrations of power—all of which, in turn, have reverse causal effects on gathering, conversion, and distribution.

I have sought to tease out the most important of these effects, and for the present, it is not necessary to go any further because the dynamics of distribution and power will first need to be examined in more detail. Again, I invite readers to explore the appendix, especially laws 2 and 3, to see where I take the ideas in this and the next chapter.

4

DISTRIBUTION DYNAMICS

DISTRIBUTION IS THE process of moving people, information, physical objects, commodities, and services among the members of a population as well as between different populations. Herbert Spencer ([1874-96] 1898) was the first sociologist to give particular emphasis to this process, arguing that as levels of production and power rise, selection for a more expanded system for moving goods, services, people, and governmental functions around a population would also escalate. Somewhat later Durkheim ([1893] 1933) emphasized the significance of communication and transportation technologies on material and moral density, which, in turn, set into motion those Darwinian-like selection pressures leading to the division of labor. And a number of more recent analyses (e.g., Boserup 1981; Lenski, Lenski, and Nolan 1991; Maryanski and Turner 1992; Mann 1986; and Hawley 1986) has similarly emphasized distribution processes as central to understanding patterns of human organization.

Yet there is a certain vagueness in these analyses, a vagueness stemming from the failure to clarify two important dimensions of distribution: (1) the movement of physical material, people, and information over space and territory, and (2) the exchange of goods and services among individuals and corporate units. Obviously the rate of exchange and the level of infrastructure for moving what is exchanged are positively related, but despite this interconnection, the transportation and communication infrastructures reveal somewhat different dynamics and consequences for a population than the modes of exchange. In this chapter, I will concentrate on modes of exchange, but before doing so, let me offer a few observations on the communication and transportation infrastructures.

The Distributive Infrastructure

In figure 4.1, the crucial dynamic processes are delineated. One causal path revolves around the levels of technology, production, and exchange. The greater the values for these processes, the more likely it is that communication and transportation networks will develop via the paths delineated in the model; and the more developed these become, the lower are what Hawley (1986) has termed "mobility costs." As mobility costs are lowered, the volume and velocity of exchanges increase, as is emphasized by the reverse causal arrow connecting mobility costs to exchange processes, and this effect will, in turn, reverberate back to increase production and raise the level of technology. The level of technology can also have a direct affect on lowering mobility costs, especially at higher levels, where the limitations of human and animal power are transcended (hence the lagged nature, or =/-, of this effect).

External trade with other populations accelerates these effects of production and exchange on the development of communication and transportation infrastructures (Braudel [1979] 1982). Trade will eventually accelerate production (hence, the lagged positive sign on this relation), while immediately increasing exchange and encouraging the development of distributive infrastructures; and in the long run, these transformations will begin to lower mobility costs.

Power processes also intersect in these causal chains, both indirectly in terms of the positively curvilinear effects of the consolidation and use of power on technology, production, and exchange (see last chapter) and directly through the expansion of communication and transportation facilities for the exercise of power. Reciprocally, a well-developed distribution infrastructure increases the consolidation of power, because it can be used to move administrative and coercive capacities about territories. External conflict accelerates these pro-

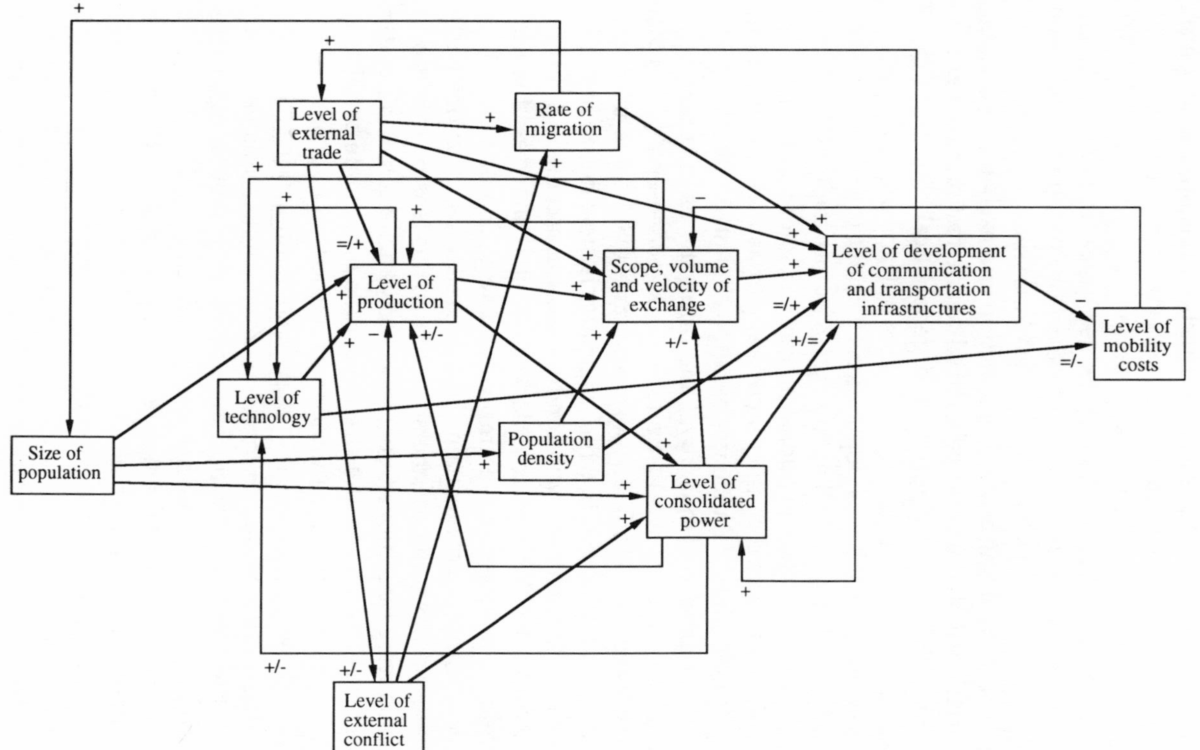

Figure 4.1. The Dynamics of Distributive Infrastructure

cesses of consolidating power and developing a distributive infrastructure, but it also has negative effects on this infrastructure via the effects of power consolidated for coercion on technology, production, exchange, and external trade. For as power becomes heavily concentrated for war and conflict, its use begins to decrease incentives for innovation, nonmilitary production, free trade, and open exchanges, thereby decreasing the development of distributive infrastructures revolving around production and exchange processes.

External trade has a positively curvilinear effect on external conflict, although specific historical and geopolitical conditions make this sign only probabilistic. Early phases of trade are often punctuated by political conflict as populations sort out the terms, spheres, and patterns of trade (Wallerstein 1974), but as trade expands and locks populations into relations of mutual interdependence, the probability of lessened conflict increases. For as exchange relations are sustained, the "balancing operations" first emphasized theoretically by Emerson (1962, 1972) and implicitly used in many empirical discussions of world trade (e.g., Keohane and Nye 1989; Keohane, Nye, and Hoffman 1993; Keohane 1984) come into play. Even unequal exchanges, in which one party is more dependent than another on the exchange, will tend to become accepted as the exchange continues. Yet as many Marxist-inspired analyses of world-system-level processes have emphasized (e.g., Frank 1979), there remains a tension over inequality in trade relations when one party dominates another; and in Emerson's terms, this tension can lead to new "balancing operations," such as coercive actions to force better terms of exchange. Thus, the effects of trade to mitigate conflicts is only a tendency that can be undone as tension mounts or as other geopolitical dynamics come into play (as is examined in chapter 6). To the degree that trade can lower external conflict, this effect can prevent the overconcentration of power that would otherwise decrease innovation, while distorting production and exchange, in ways that hamper the full range of development of the distributive infrastructure. But again, it should be emphasized that the reverse has, historically, been more often the case: concentrations of power have often biased infrastructural development toward coercive ends, and these ends often undermine the effects of external trade on such infrastructural development.

Population dynamics also feed into the development of the distributive infrastructure. Population processes indirectly influence the development of this infrastructure through their effects on the production and power dynamics examined in chapter 2. A growing population increases the logistical loads in ways selecting for increased production, regulation, and exchange; and these processes, in turn, encourage the development of the distributive infrastructure in ways discussed above. More directly, a growing population increases population density, typically in a manner that expands the division of labor (as Durkheim and Boserup argued), which, then, increases the level of exchange and movement of people in patterns fostering development of the distributive infrastructure. Similarly, high rates of migration stemming from trade or even the aftermath of war operate to increase the level of distributive infrastructural development.

In sum, then, the distributive infrastructure is intimately connected to macrodynamics. On the one hand, the distributive infrastructure flows out of the forces that population, production, power, and exchange set into motion while, on the other hand, it constrains or accelerates these forces. Having said this, the distributive infrastructure is not as likely as exchange distribution to alter fundamentally the dynamics of population, power, and production. Thus, from a theoretical point of view, the process of exchange within and between populations becomes a driving force in macrodynamics; and as exchange distribution increases, it begins to surpass other forces, such as population and production dynamics, in its causal effects on the organization of a society. It is for this reason that I separated it from gathering and producing processes.

Exchange Processes as a Driving Force of Macrodynamics

Randall Collins (1990) has proposed that "market dynamics are the engine of historical change." Earlier, Fernand Braudel ([1979] 1982) had implicitly made a similar argument in his analysis of the effects of trade and commerce on the transformation of European societies.

And still earlier, Georg Simmel ([1907] 1978) and Max Weber ([1922] 1978) weighed the significance of money and markets on the structure of society. To build my argument, I will begin by summarizing these points of view, with an eye to what each offers to a more general conceptualization of exchange distribution and macrodynamics.[1]

Weber and Simmel on Money and Markets

Weber's Model

Weber's chapter in *Economy and Society* ([1922] 1978) on "the logical categories of economic action" also contains a dynamic model (Turner 1991). For Weber, the ratio of money to nonmonied media of exchange is the critical causal force behind the transformation of market distribution and patterns of social organization. For money allows the development of credit mechanisms which pull capital and property out of local circulation networks and make them available for investment in new technologies and productive activities while facilitating the precise quantitative calculation of utilities. These changes, in turn, accelerate (1) the scale and scope of political organization (because money calculations streamline and rationalize tax collection and, hence, the capacity to support the state), which then leads to efforts at rational as opposed to nonrational regulation of productive units, (2) the volume and velocity of exchanges in markets, which also increase in the ratio of rational (profit-making) to nonrational (tradition, status-oriented) productive units, and (3) the individualization of needs, which increases the diversity of demand in markets. All of these three outcomes of money are interrelated, and after a certain threshold, their mutually self-escalating effects are irreversible. Moreover, these forces are a change-producing juggernaut which accelerates the level of production, the extension of markets, and the creation of ever more individualized tastes and needs. In figure 4.2, these processes are delineated.

I have obviously extracted and abstracted from Weber in constructing this model and added signs to the arrows that are consistent, I believe, with Weber's analysis. What is immediately evident is that the signs are all positive, underscoring the conclusion that once

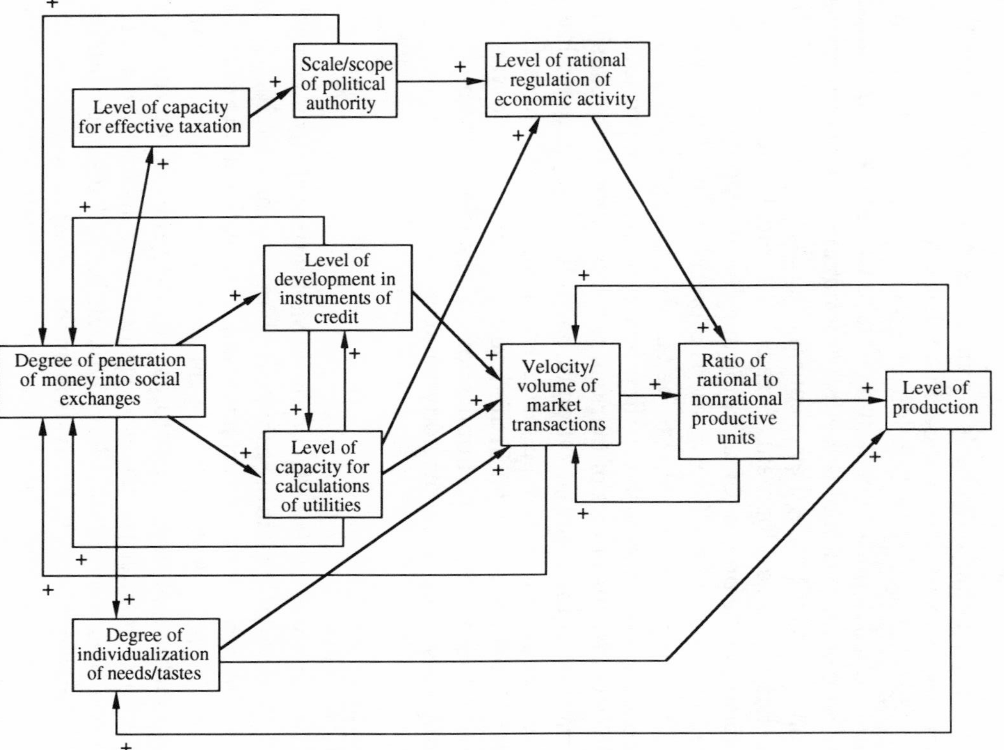

Figure 4.2. Weber on Markets, Money, and Exchange

money penetrates social exchanges at some threshold level, forces promoting social transformation (and, for Weber, rationalization) are difficult to stop, especially because of the positive reverse causal chains in the model.

The critical points in Weber's argument are these: increases in the volume and velocity of exchanges in markets are both the outcome of money, production, consolidation of power, rational calculations, and individualization of needs and tastes as well as the central cause behind the acceleration of these transformations. For once markets exist on a sufficient scale, they encourage the use of money, which, in turn, enables the development of credit instruments and rationality; they facilitate the process of usurping surplus to support the further consolidation of power (especially as a precise value can be established for the property that is to be taxed), while providing a new source of wealth themselves; they enable expanded production, which is fed by the effects of markets and money on the expansion, elaboration, and individualization of tastes (because money allows for the expression of diverse tastes and liberates exchanges from barter). Weber did not go far in this line of argument, but he was one of the first to delineate the implications of changes in the mode of exchange distribution on macrodynamics.

Simmel's Model

Simmel's analysis in *The Philosophy of Money* ([1907] 1978) compliments Weber's by providing more detail on how the introduction of money alters the nature of exchange and, in turn, the nature of social relations. In contrast, Weber's account provides more insight into the transformations of macro structures (i.e., the state and production) connected to market processes. Together, their analyses present a more complete picture of a key dynamo behind macrodynamics.

The thrust of Simmel's analysis is on the consequences of monied exchanges on the form of social relations and, somewhat indirectly, on macro structures. As with many other theorists of his time, Simmel was concerned with the question of the individual's attachment to groups in the face of growing size, rationalization, differentiation, and "objectification" of the social universe. Thus, as more impersonal standards of discourse and media of exchange—e.g., intellect, logic, law, and money—are inserted into social relations, there is an increase in personal freedom and autonomy at the expense of the more emotional and enduring attachments provided by tradition, religion, custom, strong bonds, habit, and other integrating forces in smaller, more simply organized populations. Increasingly money encourages rational calculations, devoid of the same level of emotion and attachment provided by cohesive groups, longstanding traditions, and particularistic beliefs. Yet as if to offer the counterargument to Weber's "iron cage" metaphor and to Durkheim's concern with anomie, Simmel also emphasized the positive functions of money, for both individuals and social structure: it frees individuals from constraint and creates more flexible structural arrangements.

In figure 4.3, Simmel's ideas are modeled in a manner corresponding to the presentation of Weber's. Like Weber, Simmel sees the penetration of money into social exchanges as increasing resources available for taxation by a centralized political authority, but Simmel says more: the widespread use of money generates pressures for its regulation through the concentration of power in order to preserve its worth and value, which, in turn, further encourage the expansion of regulatory authority (Simmel [1907] 1978:171–84).

Furthermore, the very existence of money, coupled with its regulation, generate a new basis for trust in social systems: the expectation that the receipt of money can be used at its stated value in future exchanges (Simmel [1907] 1978:177-78). Such implicit trust reinforces individual commitment to a social system and constitutes an important legitimating mechanism for political authority. Conversely, inflation destroys trust and rapidly delegitimates political authority.

Other integrating processes revolve around the lower density but increased multiplicity (Simmel [1907] 1978:307) and extensiveness (ibid.:180-86) of ties that result from the increased velocity of exchanges—an early version, I think, of Mark Granovetter's (1973) "strength of weak ties" argument. For as ties can cover more network space and criss-cross among groupings, a powerful integrative force is set into motion, especially when trust in the worth of money can be sustained. Moreover, high-velocity exchanges increase rates of interaction; and hence, such high velocity exchanges promote integration by enabling contacts with diverse others and by facilitating repeated interactions (Simmel [1907] 1978:292).

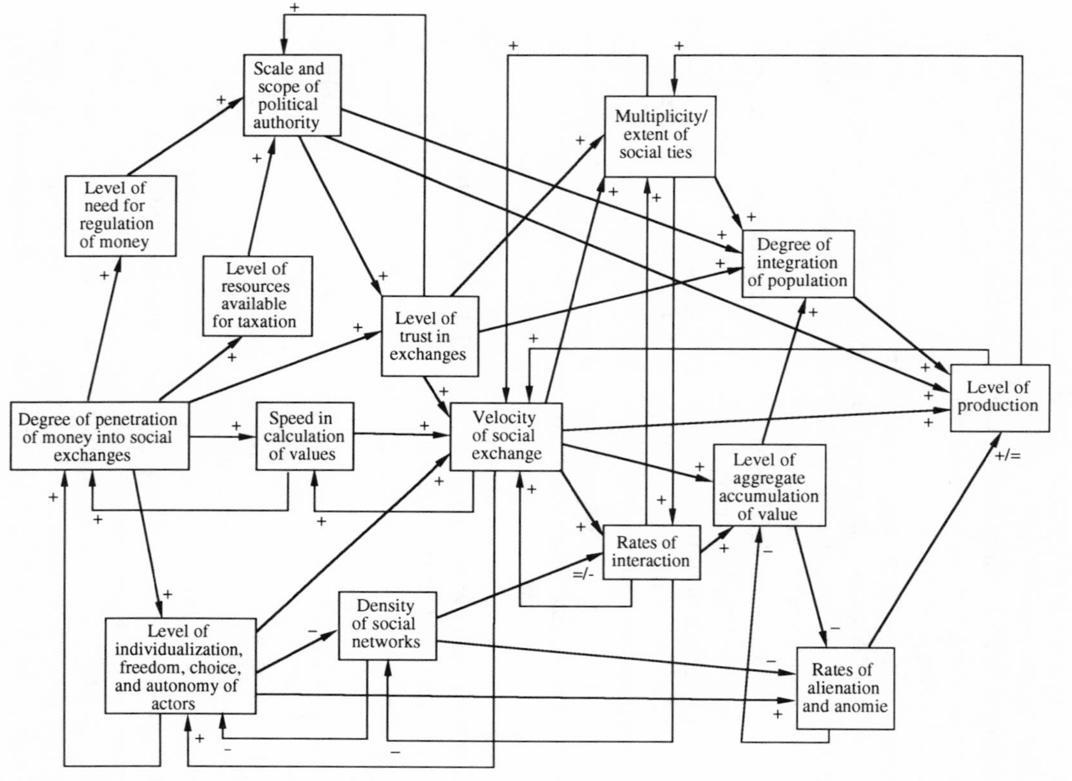

Figure 4.3. Simmel on Markets, Money, and Exchange

High-velocity exchanges also allow for the accumulation of value, because if individuals exchange resources they must (in Simmel's eye) be receiving value, and if they can increase the number of exchanges they must be accumulating ever more value. As a consequence, the more value that is accumulated, the more committed members of a population become to the macro structure and culture organizing their daily routines.

In enabling the centralization of political authority, and at the same time creating pressures on such authority to protect its worth and value, money creates regulatory authority that has an interest in sustaining the trust and legitimacy engendered by stable sources of money and, indirectly, in promoting multiple and extensive ties as they create bridges and high rates of interaction as these promote the accumulation of value. This integration may not be as cohesive and consensual in systems using nonneutral media of exchange, but it is nonetheless integration and probably of a more flexible variety, and it provides a basis of organization that can allow for population growth and increased production.

At the level of the individual, Simmel, like Weber, sees money as causing individualization, giving people more freedom, autonomy, and options. In turn, such individualization decreases network densities while increasing the velocity of exchanges, which then increases rates of interaction, multiplicity and extent of ties, and accumulation of value (Simmel [1907] 1978:307-27). Yet at the same time, individualization lessens the emotional quality and intimacy of attachments (ibid.:454), which deny individuals a major source of accumulated value while increasing the distance between self and the objects of self-expression which are too easily bought, sold, and discarded (ibid.:297). Moreover, individualization promotes alienation by virtue of fragmenting interactions (ibid.:454) and anomie by lifting constraints and barriers on perceptions about what can be acquired (with just more money).

In sum, we can take several important theoretical lines from Simmel's argument: money in markets is not only a more efficient vehicle for surplus extraction by centers of power but also a force demanding its regulation as well as a source of trust in social relations and legitimacy for centers of power (conversely, when unregulated, as is the case with rapid inflation, money decreases surplus extraction, trust, and legitimacy); money in markets increases freedom, choice, and autonomy, which, in turn, accelerate the use of money and market dynamism while providing alternative sources of value to members of a population and, at the same time, yet another source of legitimacy for central power (conversely, money can become a delegitimating force if exchanges are, as Marx [(1867) 1967] predicted, exploitive); and money in markets fundamentally alters the basis of integration in societies toward less-dense networks of weak, criss-crossing ties where value can be readily accumulated through the capacity to realize needs and tastes with the use of money in markets.

Braudel on Markets

As part of his larger project on the history of material life in Europe, Fernand Braudel (1977, [1979] 1982) has viewed markets as a transforming dynamic. In all of his work on markets and commerce, Braudel distinguishes between "lower-" and "upper-" level markets. At the lower level are found (1) person-to-person barter, (2) person-to-person exchanges using money, (3) peddlers who personally make goods, who sell them for money, and who extend credit, and (4) shopkeepers who sell goods that they do not make for money and on credit. At the upper level are found (5) fairs or relatively stable geographical places where higher volumes of goods are bought and sold with money and on credit by numerous sellers, (6) trade centers or permanent locations, where brokers and bourgeoisie sell goods and services, including credit and other financial instruments, and (7) private markets, where merchants engage in high-risk and high-profit speculations in trade involving long chains of exchange between producers and buyers.

For Braudel, the European economy became more advanced than the rest of the world's economies because of the operation of (6) and (7) in upper-level markets that were not overly regulated by state power. Thus, borrowing a page from Weber, Braudel argues that the coexistence of a comparatively nonintrusive state with a viable bourgeoisie, a system of brokers, and an efficient set of credit mechanisms enabled speculation in long chains of buying and selling which were unique to Europe and, eventually, led to the emergence of capitalism.

As an important footnote to this conclusion, I should stress that for Weber and Braudel as well as others who have followed their leads (e.g., Moore 1966; J. Hall 1985; Mann 1986; Wallerstein 1974), the existence of a nonintrusive state enabling capitalism to emerge was the result of a confluence of historical events: the rise of urban areas populated by bourgeoisie, the partial democratization of power in response to demands by nobles and the bourgeoisie, and the participation of some landholding nobles in commerce. Under these conditions, the state did not seek to dominate long-distance trade and the formation of upper markets.

Much of this upper-market activity did not initially affect the great mass of the population, because the latter's daily economic life was conducted at lower market levels. Yet the large profits of the upper market, especially those from long-distance trade and speculation, created the capital that could be further inserted into the economy. As such capital accumulation occurred, its affects reverberated down a market pyramid consisting of a few elite merchants at the top, through the more numerous bourgeoisie and shopkeepers, to the many peddlers and barterers operating among the masses. Thus, as these forces unfold, the volume and velocity of markets using money and credit penetrate ever more spheres of social life, thereby transforming the organization of a population. Moreover, to infer a conclusion pursued by Collins (1990) in Braudel's analysis, high speculation/profit markets involving long chains of exchange in terms of credit and other financial instruments (stocks, bonds, futures, etc.) are vulnerable to crisis, and even collapse, with the result that their transforming effects have often come from the social disruption and dislocation that such market crises can generate.

These transformations—whether from market growth or collapse—were not possible, however, without the development of a "national economy" in which the state (1) consolidates domestic territories and markets into a coherent system of trade, (2) encourages through its own geopolitical activities the articulation of the national system of production and market distribution to regions beyond national boundaries, and (3) resists tendencies to usurp most of the surplus capital for its own narrow political needs, interests, and privilege. For Braudel, England by the early 1800s was the first nation to meet these three conditions.

Contained in Braudel's rich description and in his own analytical statements is a more general model of market dynamics. This model is delineated in figure 4.4, although I have added additional causal chains that are consistent, I think, with Braudel's intent. As the level of production increases, pressure for the expansion of markets and the development of political power intensify. But these effects are lagged (hence, the =/+ sign on the causal arrows) because lower-level markets are adequate until production reaches high values and because the concentration of power requires substantial surplus wealth and property to support a separate administrative-coercive state. But once these lagged effects pass the necessary threshold, the volume and velocity of market exchange increase, encouraging the widespread use of money and, in turn, the development of credit mechanisms and financial instruments. As a consequence of these events, the length of the transaction chains between consumers and producers can be extended by merchants operating in restricted markets, and eventually the pyramiding of markets can occur with the top level eventually setting the parameters for transactions at the lower level. For example, markets for credit, futures, money, insurance, bonds, stocks, and so on increasingly circumscribe the dynamics of markets lower in the pyramid.

This expansion of transaction chains and pyramiding can occur, however, only when the concentration of power reveals a particular configuration: a combination of coercive and administrative control of extended territories, but the critical condition is the encouragement of comparatively unregulated long-distance trade across territories. To a degree, convergent and contingent historical events must exist to create conditions where the state is sufficiently strong to protect those engaged in long-distance trade and, yet, not so strong as to intervene and usurp profits from such trade. One historical path is for successful merchants to create a coercive wing in the state to sustain commerce, as was the case with the merchants of Venice. Another path is for a conquest state to control borders of territories but leave trade routes intact while not taxing excessively the profits and property of those who begin to create ever larger trade routes.

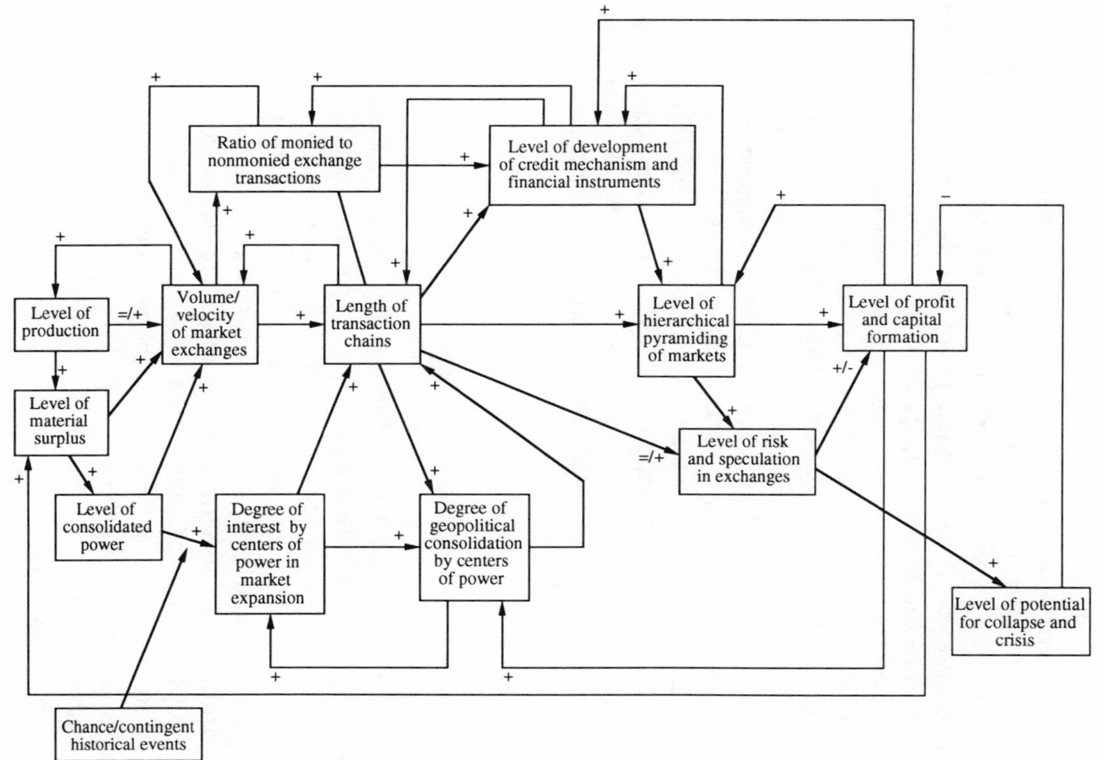

Figure 4.4. Braudel on Markets and Macrodynamics

Once initiated by any number of historical scenarios, this set of processes is mutually reinforcing. As the scale and scope of markets increase under favorable geopolitical conditions, longer transaction chains revolving around speculation with credit and other financial instruments ensue, as does the pyramiding of markets. These processes, in turn, increase profits and capital, which are fed back into markets, thereby encouraging production. Such capital formation is also in the interests of concentrated power which draws upon profits in markets, as these encourage production and further market expansion, to sustain and expand itself. Such expansion of the state can work to promote conditions of escalated market activity within a society and with other societies. Alternatively, the state can begin to coercively control and regulate territories in ways that reduce market activity. To anticipate Collins, such coercive control is especially likely when the level of risks in market speculation leads to upper-market crisis and collapse, which, then, decrease capital in ways that reverberate down the pyramid of markets. Under these conditions, political authority may begin to intervene coercively in market processes in order to extract resources to support itself and, as a consequence, accelerate the collapse of the market.

Collins on Market Dynamics as the Engine of Change

Drawing on Weber and Braudel, while critiquing Marx, Randall Collins (1990) has emphasized that market expansion and differentiation transform production and, hence, macro structure. At the same time, such expansion and differentiation can also create a series of crises and potential collapse, which transform macro structure even more dramatically, ushering in new modes of exchange and new forms of property. For Collins, then, crises in markets are the "engine" of social change, a line of argument paralleled by some Marxist analyses (e.g., Baran and Sweezy 1966).

In figure 4.5, Collins's general line of argument is modeled. Following Braudel, Collins distinguishes between two levels of markets, (1) those that are local and face-to-face in which custom, tradition, and interpersonal surveillance and sanctioning enforce fair prices, low profits, and reduced exploitation; and (2) those that are

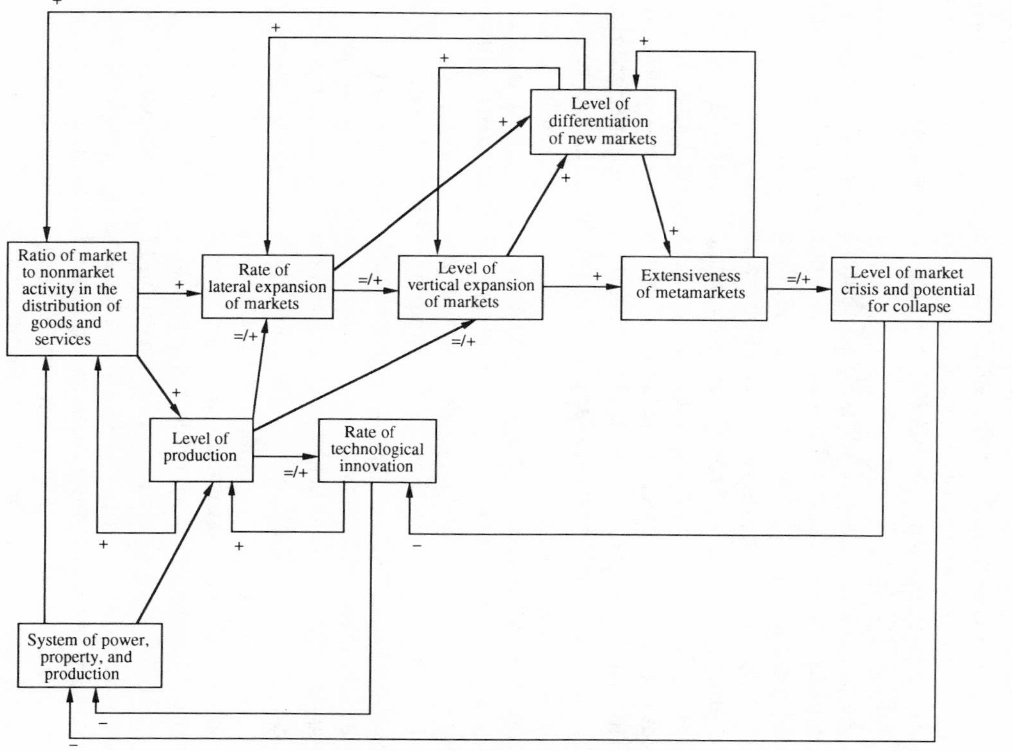

Figure 4.5. Collins on Markets as the Engine of Macrodynamic Change

long-distance and future oriented in which speculators and brokers manipulate exchanges in order to achieve large profits. These second-level markets are viewed as "superordinate," involving some degree of monopoly or oligopoly to keep competition from driving prices and profits down as Marx had argued in his scenario about the falling rate of profit for capitalists. Such markets thus allow for very high prices and profits but not so much as to strangle the demand for goods.

There are several critical dynamics inherent in these superordinate markets. As figure 4.5 delineates, they tend, first of all, to expand laterally across territories and, in so doing, they connect large numbers of people and productive units into networks of exchange. Secondly, after a certain degree of lateral expansion, markets begin to expand vertically in the sense that the medium of exchange in one market becomes the object of exchange in a new, higher-order market. Thus, for example, money as the medium of exchange becomes, itself, subject to market manipulation. Once initiated, such pyramiding can take many forms, as when futures, options, bonds, stocks, insurance premiums, mortgages, and other media facilitating exchange in one market become, themselves, the objects that are exchanged in new markets. These markets are "metamarkets" in Collins's terms because they exist above and beyond more local markets and involve high risk and speculation. And although they facilitate the creation of new types of markets, they also are subject to crises and collapse. And when such collapse occurs, incentives for technological innovation are reduced and new forms of property, production, and power are likely to emerge.

Much of Collins's argument is directed at providing an alternative to Marxist interpretations of change, and so the vocabulary and thrust of the argument do not fully correspond to my purposes here. But there are several important leads in his analysis: markets have the capacity for self-transformation, independently of production and power dynamics; such transformations revolve around lateral expansion, vertical pyramiding, differentiation of new markets, and metaorganization of high-risk and high-speculation markets; crises and collapse of metamarkets have significant effects on the organization of production and power dynamics.

A General Model of Distributive Exchange Dynamics

This diverse group of thinkers, along with many others who have made the same points, provide most of the necessary elements for a general model on the effects of exchange dynamics on macro-level social processes. In figure 4.6, the contours of my argument are outlined.

The chain of causally connected forces—beginning with the level of technology and moving straight across the middle of the model to level of potential for crisis and collapse in markets—is a good place to initiate a review of exchange distribution. And, as we move along, the effects of other forces in the model can be introduced. As we have seen in chapter 3, production and technology are positively related in a mutually reinforcing cycle. This cycle is, as noted in chapters 2 and 3, affected by the size and growth rate of a population in a positively curvilinear pattern, initially causing a stimulus to increased production and, indirectly through the reverse causal chain from production to technology, to the rate of innovation. Yet eventually, population size operates as a drain on production as very large numbers of people begin to consume surplus and, thereby, deny capital and entrepreneurial incentives to the productive process. Power also feeds into the production process in a positively curvilinear way, encouraging production in order to generate the surplus needed for concentrating and sustaining power, but as very high degrees of concentrated power are realized, the use of power discourages incentives for technological innovation, consumes surplus physical capital, and overregulates entrepreneurial activity. As technology, population and power affect production, pressures for expanding markets as a mechanism for distributing goods and services increase.

These pressure are relatively low until a threshold of production is reached, revolving around nonkin entrepreneurial structures organizing nonhuman sources of power with differentiated human labor and capital to produce goods for consumption by others (hence, the lagged positive, or =/+, relation between production and markets). Braudel's distinctions within lower markets capture the initial phases of a developmental sequence in response to varying levels of production—person-to-person barter, person-to-person exchange

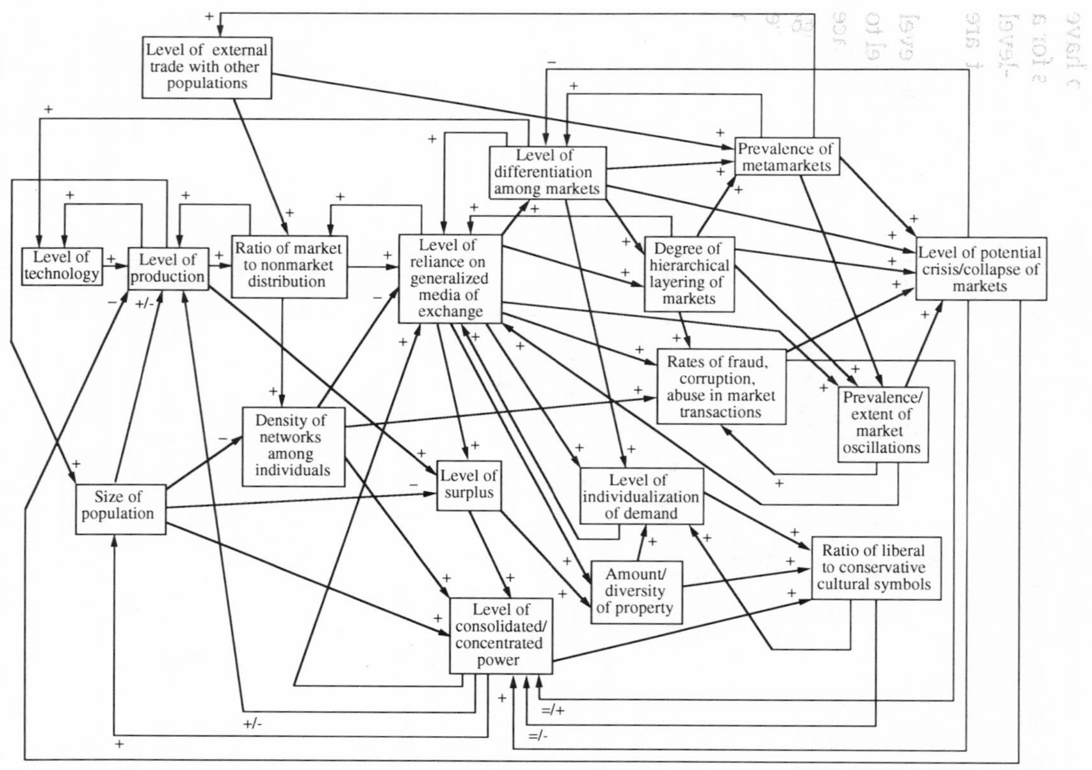

Figure 4.6. The Dynamics of Exchange Distribution

with money, peddlers who make and sell goods while often extending credit, and shopkeepers who sell goods that they do not make for money and on credit. Braudel's portrayal of upper markets continues the developmental sequence—fairs, trade centers, and eventually long-distance trade and speculative private markets. As the latter stages of Braudel's sequence indicate, external trade with other populations is an especially strong stimulus to increasing the ratio of market to nonmarket distribution, a stimulus that feeds back to increase production and technology.

As is also evident, as markets develop, generalized media of exchange such as money and credit are increasingly used; and the more they are used, the more they stimulate the development and expansion of markets because money and credit pull resources out of traditional patterns of use and consumption, freeing resources so that they can be used in new locales and in new combinations. Moreover, with established and stable media of exchange, markets become more differentiated as a result of the capacity for money and other media to "individualize" demand and, thereby, to encourage new combinations of production. As long as exchange is tied to barter, to nonmarket transactions, and to local markets where resources are limited, the preferences of individuals are constrained and the range of exchange is limited. With money and other financial instruments, individuals have the capacity to express preferences with the buying power of a neutral medium; and once this capacity is established, a mutually reinforcing cycle among individualized demand, reliance on generalized media, and market differentiation to meet ever more diverse demand ensues.

Such individualization of demand is also facilitated by demographic forces. As populations grow, the density of ties among members of the population lessens, a force which reduces the power of informal sanctions but which, at the same time, fosters greater reliance of generalized media to mediate new patterns of social relations and new options for individuals. Moreover, through its indirect effects on increasing the ratio of market to nonmarket transactions, generalized media operate to lower network density, as Simmel implied, and to increase reliance on these media and, hence, to further individualize taste and choices, which are translated into more diverse market demands. The same set of processes, however, also increases logistical loads for more formal control and regulation, thereby increasing Spencerian selection for increased concentration of power.

The dynamics of power are further involved in this cycle in an even more fundamental way: the use of money in markets creates yet another source of surplus wealth that can be used to support the consolidation and use of power; and as a result, centers of power have a vested interest in maintaining the viability of this source of surplus, with the result that they inevitably take control of the money supply. Yet as we will see in the next chapter, this control almost always causes problems, as the centers of power (*a*) issue too much money and credit (and thereby cause rapid inflation), (*b*) usurp too much surplus revenue and property (thus, causing disincentives for production as well as incentives to revolt), and (*c*) fail to appreciate the extent to which their legitimacy in a market driven system is dependent upon the stability and trustworthiness of the media of exchange. But to the degree that centers of power can keep from becoming too concentrated and can resist overtaxation of wealth generated in markets, overprinting of money, and overextension of credit, these centers have positive effects on increasing the reliance of the population on generalized media and, through the reverse causal effects of this process, on increasing penetration of markets and the resulting expansion of production and technological innovation. Rarely do centers of power achieve this balancing act between over- and underregulation, however. As a consequence, delegitimation of power and disintegration of the population eventually occur.

Another very important effect of individualization of demand on power, money, and market differentiation occurs through the increasing ratio of liberal to conservative symbols, a critical force recognized by Spencer ([1874–96] 1898) and later Pareto ([1916] 1935). Symbols emphasizing freedom, choice, options, self-determination, and restraints on external power inevitably increase in frequency as money, markets, individualization of demand, and diversification of ever greater amounts of objects defined as property all operate in a set of mutually reinforcing cycles. These symbols need not be initiated by market processes, however, for they may often have their origins in other processes, usually in religious, social, and political movements. Whatever their origin, market forces make them ever more salient; and such liberal symbols further legitimate the rights of individuals to reveal their own preferences and, at the same time, constrain centers of power from overuse of their regulatory authority and coercive powers (indeed, these symbols are used to mobilize demands for increased democracy in the selection of decision makers in the polity). The negative effect on overuse of power is, I hypothesize, lagged in that the ratio of liberal to conservative symbols must reach high levels before constraining political authority in ways that threaten its legitimacy and viability should these symbols be violated. But in constraining power these symbols work indirectly to encourage market development and differentiation mediated by restrained taxation of surplus and by efforts of centers of power to maintain stability in money, credit, and other financial instruments.

Once these forces have created reliance on generalized media, market differentiation, and individualized demand, additional forces are unleashed. One such force is more frequent market oscillations in which overproduction relative to demand initiates recessionary processes (in which production and demand decline). All markets are subject to these oscillations, but the more money and other generalized media are used, the more markets and, in turn, gathering-producing become susceptible to these "business cycles." These cycles become even more likely with differentiation, hierarchical layering, and metamarkets because, with recessionary downturns, overspeculation in metamarkets can be exposed as demand and production decline. The result is for the downturn to become more rapid and, potentially, deeper; and when there has been rampant overspeculation—as was the case in the Spanish bankruptcies of the 1500s, the South Sea Bubble in France, or the 1920s in America—normal market oscillations can lead to crisis and collapse. Thus, market oscillations between high demand—high production and low demand—decreased production become typical of developed markets; and they can also become the opening step in crisis and collapse in markets in which overspeculation in the terms and instruments of exchange has occurred. Such collapse requires considerable layering of markets into hierarchies as both Braudel and Collins emphasize.

More Marxist-inspired arguments would not dispute this emphasis on market hierarchies, but would stress that the potential for collapse inheres in all highly competitive markets regardless of the level of hierarchy. In their view, market competition forces a "falling rate of profit" (Marx [1867] 1967; Appelbaum 1978) as producers get locked into cycles of introducing labor-saving technologies which reduce labor costs but which are soon copied by other competitors, thereby negating the competitive price advantage temporarily enjoyed by the introduction of such technologies. And as these technologies reduce the labor force, they also take income from workers with the result that aggregate demand in the market declines, placing further pressure for price reduction through cost savings and, thereby, locking capitalists into a self-destructive cycle. Thus, oscillation, crisis, and collapse are, from a more Marxist perspective, built into all competitive markets. But I think that Collins is correct in the view that real collapse of markets is far more likely with high risk speculation in layered markets than in tendencies for falling rates of profit. Indeed, key actors in markets are far more likely to form networks to stave off destructive competition, and in so doing, they often contribute to overspeculation by removing productive activities from the "discipline" of markets, for as these noncompetitive networks are sustained they are increasingly used to speculate in areas not connected to production of actual products.

Once some degree of differentiation of markets using generalized media exists, hierarchies among markets begin to emerge, with those higher in the hierarchy ultimately constraining those at lower levels. The hierarchy is established in terms of (*a*) the distance and scope of transactions, (*b*) the use of generalized media, (*c*) the control of transactions by a limited number of actors, and (*d*) the level of speculation involved in transactions. That is, the greater is the geographical scope, reliance on diverse generalized media, control of exchanges by a limited number of parties, and speculation over generalized media, then the higher in the hierarchy is a market. Thus, although local markets lower in the hierarchy can operate with considerable autonomy, once long-distance trade among limited numbers of actors (e.g., cartels, oligopolies, chartered corporations) using ever more sophisticated financial instruments (e.g., money, credit, futures, options, etc.) and engaging in speculation over the media of trade come into existence, they tend to expand; and over the long run, the high profits in such markets attract capital, entrepreneurial activity, and regulation by political authority. As a result, they come to dominate lower markets, in several senses: (*a*) they increase the use of money, credit, and other financial instruments in lower markets; (*b*) they build up new market infrastructures, such as banking, wholesaling, and insuring, that begin to affect trade at lower market levels; (*c*) they create great centers of wealth and property, especially when controlled by a limited number of actors, which translate into demand for luxury goods and crafts in local markets; (*d*) they stimulate the development of distributive infrastructures (roads, harbors, ships, airplanes, canals, trade centers, warehouses, etc.), and most importantly, (*e*) they create the potential for market collapse as a result of overspeculation, with such collapse reverberating through lower-level markets.

This last effect of higher-level markets is more likely when these kinds of markets become, in Collins's terms, "metamarkets" dealing in the media and terms of exchange in other markets. Thus, if money, insurance premiums, mortgages, options, futures on yet to be produced goods, stocks, bonds, and other instruments facilitating the operation of other markets become, in themselves, differentiated markets, the high profits to be made from speculation in these markets (since the discipline imposed by manufacturing costs and time in actually making, storing, transporting, and selling a physical commodity is attenuated) leads to overspeculation and, eventually, a crisis and potentially a collapse of the market.

Such potential is dramatically escalated by the propensity of all higher-order markets using complex financial instruments to reveal fraud, corruption, and other abuses that hasten collapse. Only when (*a*) dense networks, (*b*) patterns revolving around secrecy, monopolistic control, and (*c*) shared cultural traditions exist among actors in metamarkets can these abuses be attenuated. Yet these forms of discipline are eventually undone by virtue of a number of forces built into the nature of higher-order markets: the very differentiation of markets loosens controls and fosters efforts by players in one market to enter others; the very nature of generalized media as ethically and culturally "neutral" gives such media the power to transcend networks and encourage speculation and investment in ways that break control of markets by a limited number of players; and the very existence of the long-distance trade on which metamarkets are ultimately built encourages geopolitical incursions by actors from other populations. Moreover, the ever escalating needs of political authority or elites for additional resources often push new actors into these controlled markets to even more dangerous speculative activity or lending practices than the original players in such markets.

Fraud and corruption in, as well as collapse of, higher-order markets are often hastened by centralized political authority, as it also becomes a speculative player in metamarkets, as it borrows and does not repay capital in money markets, as it fails to protect cartels, monopolies, and charters, as it generates patterns of elite corruption draining capital from markets, and as it disrupts agreement among actors in markets for narrow and short-term partisan or alliance considerations. Yet ironically, whether it engenders fraud, corruption, abuse, and collapse or simply must deal with the consequences of these events, political authority is always pulled into market dynamics. As the reverse causal arrows indicate, market abuse, crisis, and collapse all exert pressure to increase the concentration of power, as centers of power seek to sustain revenue flows from markets and as they attempt to control the social turmoil associated with disruption of lower-level markets, especially as this disruption aggravates tensions revolving around inequalities. And as power becomes ever more concentrated to deal with these problems, it extracts more surplus and begins to overregulate production in ways hastening collapse, or revolt.

If metamarkets collapse, this event will, for a time, decrease both the hierarchy and differentiation among markets. And if the collapse of metamarkets also brings down a political regime, then the stability, trustworthiness, and viability of generalized media are undermined, a situation that can decrease reliance on such media and, in the process, lower the ratio of market to nonmarket exchanges, levels of production, and rates of technological innovation. In turn, dissolution of the population may occur, unless new centers of power can emerge or come in from the outside to restore order and reestablish trustworthy and viable generalized media of exchange.

Elementary Principles of Distribution Dynamics

Distributive processes have enormous consequences for the size and movement of a population, the level of production, and the concentration of power; and it is for this reason that I have separated them as a distinctive force behind macro-level social organization. Populations cannot grow or move, production cannot expand, and power cannot be effectively consolidated or used without (1) the development of a distributive infrastructure for transportation and communication, and (2) the establishment of systems that increase the scope, volume, and velocity of exchanges for goods and services. Conversely, the collapse of a distributive infrastructure or the modes of exchange will cause levels of production to decrease, centers of power to destabilize, and populations to dissolve. The importance of these distributive processes on macrodynamics will become increasingly evident as we proceed. For the present, let me summarize in broad strokes the line of argument presented in this chapter in propositional form:

I. The size of a population, the level of production, and the concentration and use of power are a positive function of the level and scale of distributive processes for that population

II. The level and scale of distribution in a population are a positive and multiplicative function of:

- A. the level of development in the communication and transportation infrastructures
- B. the scope, volume, and velocity of exchange processes

III. The development of communication and transportation infrastructures is a positive function of:

- A. the scope, volume, and velocity of exchange processes
- B. the rate of migration
- C. the density of population settlements
- D. the level of consolidated power

IV. The scope, volume, and velocity of exchange processes are a positive function of the ratio of market to nonmarket transactions among the members of a population, which, in turn, is a positive and additive function of:

A. the level of production, which is:
   1. a positive function of the level of technology
   2. a positive curvilinear function of population size
   3. a positive curvilinear function of concentrated political power
   4. a negative function of market downturns, crises, and potential collapse
B. the reliance on generalized media of exchange, which is:
   1. a negative function of density of networks connecting members of a population
   2. a positive function of market differentiation
   3. a positive function of market layering into hierarchies
   4. a positive function of individualization of demand
   5. a positive function of moderate concentrations of power
C. the level of external trade with other populations

5

POWER DYNAMICS

THUS FAR, I HAVE used phrases such as "consolidation of power," "concentration of power," "centralization of power," and "the use of power" in a connotative manner. Now, it is time to give these phrases a more precise meaning and explore their operation in human populations. There are, of course, many efforts to define and conceptualize the very elusive properties of power;[1] and for macro analytic purposes, we can stay within most of these efforts by defining power as the capacity to regulate and control the actions of other members of a population and the structural units organizing these members. By regulation and control, I simply mean that one actor, or set of actors, has the capacity to affect the course of actions of other actors, or set of actors; and the greater this capacity, the more power an actor or set of actors possesses.

Conceptualizing Macro Structural Power

In analyzing this regulatory capacity for macro-level phenomena, two dimensions of power become relevant: (1) the *degree of control and regulation* of an average actor's activities by another actor and (2) *the distribution of control and regulation* among actors. The first consideration concerns the extent to which the average or modal actor in a population, including both individual and collective actors, must adhere to directives by other actors or suffer negative sanctions. The second issue deals with the degree of concentration or dispersal of control and regulation across the actors comprising a population. Figure 5.1 plots these two dimensions of power, with some illustrative referents to some general types of power that have been found historically in the organization of human populations.

The arrows flowing out of each type indicate the path of transformation for each configuration of power. For example, a warlord system in which power is dispersed among a number of actors, with each of these actors exercising high degrees of control over their constituents, tends to move toward a more centralized system of power, with one lord eventually controlling the others, at least for a time. Depending on the productive technology, this can be toward a totalitarian state (e.g., communist China), a confederation of chiefdoms (e.g., much of Africa before colonialization), a feudal monarchy (Europe after the Dark Ages), or it can, and usually does in the long run, lead to disintegration of a population as a coherent whole. A system of chiefdoms often collapses into feuding warlords, or evolves into a feudal monarchy when one can attain paramount power. A totalitarian state eventually collapses into a system of warlords (e.g., Yugoslavia in the early 1990s). A feudal monarchy usually falls back to a decentralized system of independent fiefdoms (a form of chiefdom in an agrarian population), only to be recentralized by a monarch, at least for a time (e.g., the history of Europe after the fall of the Roman Empire up to the period of industrialization). Coercive empires always seek to regulate to a high degree, but the diversity of conquered populations, the size of territory, and the logistics of regulation always prove difficult, and eventually they collapse, even as they try to increase their regulatory control (e.g., the Soviet Union). Co-optive empires have concentrated centers of power, but do not regulate to a high degree (e.g., the Roman Empire), but they too collapse, usually after erosion of the centers regulating control in outlying regions. Liberal democracies move toward more concentrated control, as the diverse needs and interests of the population create pressures for centers of power to respond to an ever growing list of citizen demands (e.g., modern Europe and the United States) or, potentially, they lose the capacity to regulate effectively as interest group politics overwhelm decision making and foster gridlock. Populist democracies usually degenerate, move to a more liberal form, or

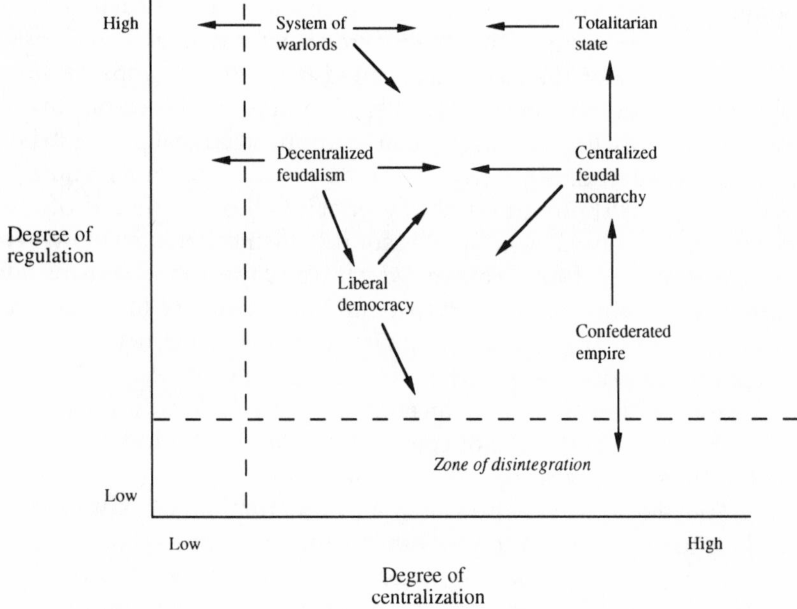

Note: Power is the capacity to regulate the activities of other social units. Its use involves a ratio of administrative to coercive control of these activities as well as the manipulation of cultural symbols and material incentives.

Figure 5.1. Variations in Macro-level Regulation and Control with Power

toward a more regulation by concentrated power (e.g., classical Greece, most "utopian communities," and all postrevolutionary situations).

My point here is that power can be either dispersed or concentrated; under either condition it can involve high or relatively low levels of control. Regulatory control and centralization can thus vary somewhat independently of each other. Take some of the extreme cases, for example: a warlord system has power dispersed across a number of often feuding warlords, but each exercises very tight control over its constituency; or a co-optive empire has a coercive and administrative center but exercises only loose control over its member regions. I emphasize this point because it is critical to understanding power dynamics at the macro structural level.

As a number of theorists has argued (e.g., Etzioni 1961; Collins 1975; Mann 1986; Blalock 1989), there are different bases of power. For example, Michael Mann (1986) has emphasized that power exists along four interrelated networks: ideological, economic, military, and political. Although I do not agree with all of the details of this portrayal, the general argument is sound: the mechanisms by which regulation and control are exercised vary; and these mechanisms fall along the dimensions listed by Mann, and by others (e.g., Collins 1975). In abstracting above Mann's and others' distinctions, four basic bases of power are evident: (1) the use of symbols, (2) the use and manipulation of material incentives, (3) the use of administrative structures, and (4) the use of coercion. Let me briefly indicate what each of these involves.

(1) Those actors with power in a population almost always use symbols to regulate and control other actors. The most effective symbols are evaluative, invoking actors' standards of what is right or wrong and what should occur; and when there is consensus over these symbols among the members of a population, centers of power can manipulate these symbols, and totems that represent them, to control and regulate. In so doing, these centers of power can also legitimate their right to engage in regulation, and with such legitimation they can pursue ever more effective manipulation of symbols for control. Yet once centers of power rely upon a set of symbols, these same symbols can be used as criteria for assessing the degree to which political decisions conform to the tenets in these symbols; and if decisions are viewed by segments of a population as not meeting the evaluative criteria of symbols, then these very same symbols can become a basis for mobilizing counterpower. Moreover, if there is a low degree of consensus over evaluative symbols or even subpopulations with conflicting sets of evaluative symbols, efforts to use evaluative symbols as a mechanism of control will simply aggravate existing cleavages. And typically, mobilization of counterpower first involves the codification of new symbols that challenge and delegitimate those of existing centers of power. Thus, the dynamics of power revolve around the use of symbols.

(2) Those actors with power are able to manipulate material incen-

tives in order to control and regulate other actors. By providing material benefits or imposing material costs on others, a considerable amount of regulation ensues. Such capacities to manipulate depend upon centers of power possessing material resources, or at least having the legitimate right (as dictated by evaluative symbols) to use the resources produced by members of a population. Moreover, centers of power must also have the administrative capacity to monitor conformity to directives and to bestow or withhold material benefits. If such capacities do not exist, as is often the case with financially strapped leaders whose legitimacy, administrative structures, and coercive capacities are weak, then efforts to manipulate incentives will often aggravate existing resentments and, in the end, delegitimate and further weaken a center of power.

(3) Those actors with power will inevitably have an administrative structure for carrying out directives, monitoring conformity, and imposing sanctions, and in so doing, these actors can control others. For once an administrative structure is in place, its existence reduces the options of others and forces them, to some degree, to work with this structure. Thus, even when there is great political turmoil, an established administrative structure can still exercise considerable control—as has often been the case, for example, in the tumultuous history of France. The more coherent an administrative structure—that is, the more clearly differentiated its offices and the authority relations among them—the greater is this capacity for control; and the more this structure is legitimated by evaluative symbols and backed by coercive structures, the greater still is its capacity for control. Conversely, if an administrative structure has ill-defined offices and relations among them, it often becomes a source of conflict and instability, thereby weakening its ability to regulate a population.

(4) Those actors with power ultimately will have some capacity for physical coercion. For if directives are not obeyed, physical force must be used if power is to be sustained. Yet to paraphrase Edmund Burke, no population can be ruled which must be perpetually conquered. Force must only be used as a last resort, or as a strategic and short-term option for particular situations; otherwise, its use generates tensions and resentments which, eventually, create countersources of power that undermine the legitimacy of a regime, the effectiveness of its administrative structures, and the hegemony of its

coercive capacities. Indeed, the frequent and pervasive use of coercion signals that the other three bases of power have not been established or have eroded away. Coercion is most effectively used infrequently and sporadically to control disruptive actions that do not signal broad-based losses in the other bases of power. In fact, the use of coercion in this more strategic manner can strengthen bases of power.

Thus, power is the ability to control and manipulate members and organizational units of a population in terms of symbols, incentives, administrative structures, and coercive capacities. Any one base of power is ineffective in the long run; power is most effective when all four bases can be brought to bear in regulatory activity—a situation that is always transitory but effective as long as it lasts. With this idea in mind, let me reintroduce two phrases that I have connotatively used in previous chapters: "the consolidation of power" and "the centralization of power."

When power is consolidated, a set of actors has been able to (*a*) use symbols, (*b*) establish a coherent administrative structure, (*c*) acquire or control sufficient resources so as to be able to manipulate material incentives, and (*d*) coerce judiciously. The less these four mechanisms are brought together, the less consolidated is power. The great problem for those seeking to use power is the ability to blend these bases in an effective combination. For example, a charismatic revolt often brings symbols and coercion together, usually with too much of the latter. And as Weber ([1922] 1978) emphasized, such a regime often has trouble establishing an administrative routine that is essential for fully consolidated power. In contrast, the political democracies of the European west and Japan have consolidated power so effectively, at least for a time, that large-scale coercion is rarely needed or used, administrative structures are coherent, manipulation of material resources is effective, and common evaluative symbols can be invoked to control actions of actors and to legitimate a political regime.

When power is centralized, the issuing of directives is concentrated in relatively few persons and offices, with other offices of the administrative structure carrying out instructions from above. Conversely, when power is decentralized, directives emanate from diversely situated offices, with less domination by a few key people and offices. As figure 5.1 denotes, there must be *some degree* of centralization of power if control and regulation are to be possible and if disintegration is to be avoided. Yet as this figure also illustrates, there is a great range in *how* centralized power is, even as regulation remains constant. For example, a system of warlords and a totalitarian state may regulate the average actor to a similar high degree, but vary enormously over the degree of centralized power. Similarly, feudal systems will tend to cycle from centralized (under a strong monarchy) to more decentralized profiles (when lords revolt or withdraw support from the monarch), but the level of control over, and regulation of, the average actor can remain roughly the same.

Although the consolidation and centralization of power can vary independently of each other, they tend to be positively correlated. For as legitimating and manipulating symbols are mobilized, as coherent administrative structures are developed, as effective use of material incentives is pursued, and as coercion is minimally brought to bear, some degree of centralization ensues, for several reasons: legitimating symbols need objects, and these will tend to be higher-level persons and offices, although lower-level offices and persons must also receive legitimacy; coherent and effective administrative structures reveal clear hierarchies of authority; manipulation of material incentives must be consistent, which implies a set of orders from a few persons and offices; and coercion must be only strategically and minimally used, which implies higher-order control of the structures of coercion. Some patterns and profiles of consolidation cause more centralization than others. For example, when power relies on coercion under conditions of weak legitimacy, the administrative structure and the manipulation of incentives will be highly centralized. Conversely, when coercion is rarely needed and legitimacy is high and consensus over evaluative symbols is also high, the administrative structures and manipulation of incentives can be much less centralized. I do not want to construct a typology on these issues, but only to emphasize that the consolidation of power produces some degree of centralization and that the degree of centralization will reflect the configuration or combination among the bases power that are brought together. As a general rule, the more frequently must coercion be used as a mechanism of control and regulation, the more centralized will be the profile of power—a point made by diverse scholars from Thucydides through Herbert Spencer to many modern commentators.

With these conceptual preliminaries, we are now in a position to address the dynamics of power which revolve around the forces affecting, and being affected by, the consolidation and centralization of regulation in a population. These forces are modeled in figure 5.2.

As would be expected from the discussion in previous chapters, one force affecting the dynamics of power is population size, another is production, and still another is distribution. I have added for special emphasis two more forces not given extensive treatment in previous chapters: (1) inequality and internal threat, and (2) external threat and conflict. Together, through the causal paths delineated in the model in figure 5.2, power dynamics play themselves out. Each of these forces, as it affects and is affected by the consolidation, centralization, and regulatory dimension of power, will be examined below.

Population and Power

As Herbert Spencer ([1874–96] 1898) was the first to conceptualize theoretically, population growth increases the size of the "social mass" and, thereby, generates logistical problems which, if unresolved, lead to the dissolution of the population. As emphasized in the previous chapters, this potential generates selection pressures for the consolidation of power in order to regulate, control, and coordinate activities of population members and the structural units organizing these members. Such consolidation involves the emergence of leadership positions which can make decisions for the population as a whole, along with the legitimating symbols, possession of material incentives, administrative structures, and coercive capacities necessary to carry out and to enforce decisions. And the larger a population gets, the greater are the selection pressures for consolidated power. If such consolidation does not occur because of low consensus over evaluative symbols, inefficient and corrupt administration, lack of material incentives to manipulate, or low coercive capacities, then the potential for dissolution of the population is greatly increased.

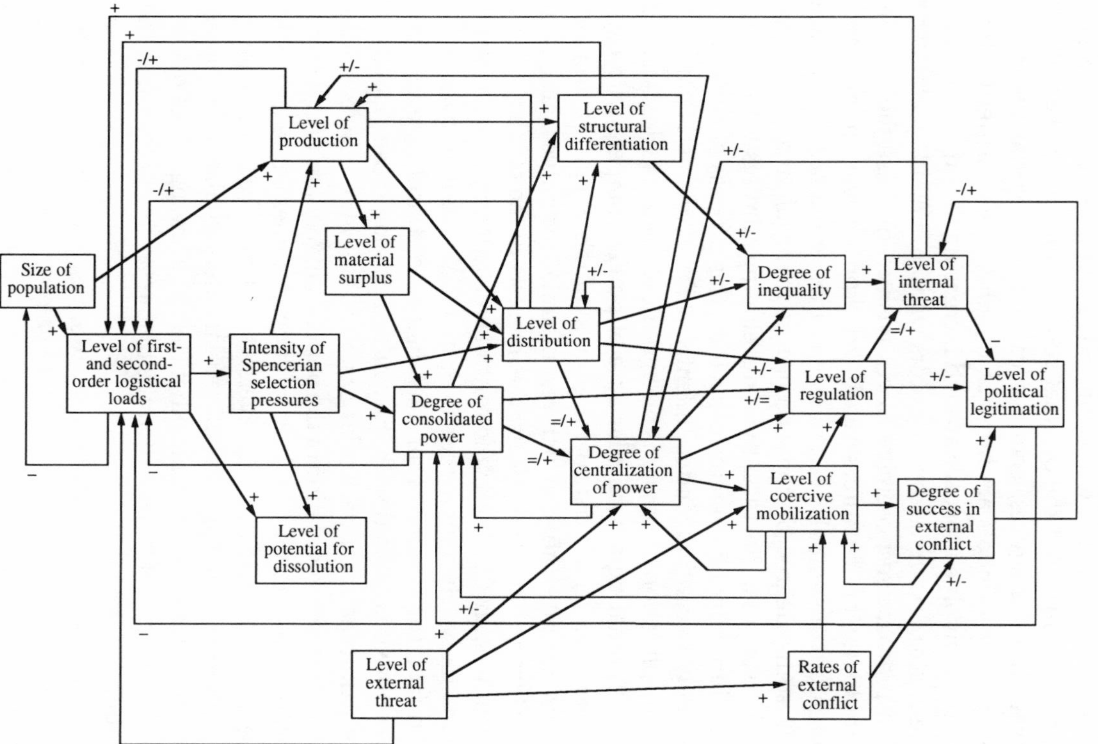

Figure 5.2. The Dynamics of Macro-level Power

The sequence of events to consolidating power can vary enormously under these selection pressures, but once the process is initiated it has a life of its own, often independently of the selection pressures that initiated the original consolidation of power. For as administrative structures, coercive capacities, material resources, and evaluative symbols are brought together, the short-term interests of those involved revolve around increasing the efficacy of each of these bases of power. Administrators want to expand their administrative prerogatives; coercive forces wish to increase their readiness; holders of resources wish to horde ever more; and manipulators of symbols seek to extend their influence. Thus, once there is some degree of convergence in manipulating evaluative symbols, using material incentives, developing administrative structures, and mobilizing coercive capacities, this process of consolidation is likely to continue independently of Spencerian selection.

Yet consolidation of power is often difficult, because the interests of those dominating a particular base of power compete with each other. For example, wielders of religious symbols often seek to create their own administrative structures, material incentives, and even coercive forces that stand against the interests of the secular state or holders of wealth. Or, those who control wealth, commerce, or property are usually reluctant to pass their material resources over to the state. In fact, the history of all movements to consolidate the four bases of power is littered with the conflicts among those disproportionately dominating a particular base; and until there is consolidation of these bases, regulation of the population remains problematic, logistical loads increase, and dissolution becomes more probable. Moreover, if consolidation is reversed and new conflicts emerge among, for example, religious and secular rulers, civilian administrators and military elites, or some other cleavage among actors operating to hold a base of power, then the disintegrative potential of the population is high.

This potential, or its actual realization, is one force behind the centralization of power. Users of power recognize that they must control all of its bases—symbols, material resources and incentives, coercive capacities, and administrative structures; and as a consequence, much of the conflict among actors operating from different bases of power is over who can grab the other bases and consolidate them under one ruler, or set of rulers. At times it is the manipulators of evaluative symbols who win out (e.g., as is the case in contemporary Iran); at other times it is the forces of coercion (e.g., virtually any nation experiencing coup de gras and military juntas); sometimes, though more rarely, it is the sources of commercial wealth (e.g., the "merchants of Venice," the "United Provinces" of sixteenth-century Holland, or the early industrialists in England and the United States) who are ascendant; and in the long run, if centralization of power is to be stable and effective, it is the civil administrators who come to dominate. Indeed, the great accomplishment of modern political democracies is that centralization of power involves administrative dominance over centers of coercion, coupled with a sharing of control over material incentives with actors outside this administrative system and an effective system of legitimating symbols that both bestows and limits the regulatory capacities of administrative and coercive bases of power. But in all cases, consolidation of power always generates pressures for the centralization and concentration of power; and this pressure was, I argue, originally caused by population pressures (Maryanski and Turner 1992; see also, Carneiro 1967, 1970, 1973; Fried 1967; Johnson and Earle 1987; and Earle 1984). And even today, population growth and the specter of Malthus's four horsemen can still drive the consolidation and centralization of power, although other forces become increasingly involved as societies become more complex and differentiated. Moreover, when production and distribution raise the normative standard of living, birthrates decline and reduce population pressures as the engine of consolidation and centralization of power, but new pressures and logistical loads arise from (*a*) the complexity of production, distribution, and even the administration of power, (*b*) the tensions associated with inequality, and (*c*) the burdening of geopolitics and external threat.

Production and Power

Population, Production, and the Dynamics of Power

Population growth often drives power dynamics indirectly, via its effects on production. For as the "social mass," to use Spencer's term, gets larger, production and eventually distribution as well must expand; and as they do so, they (*a*) create the surplus wealth necessary to sustain centers of power, (*b*) generate logistical and entrepreneurial problems (e.g., over transportation, communication, monetary systems, market abuses, and the like) whose management requires independent sources of power, and (*c*) aggravate tension- and conflict-producing inequalities that require centers of power to control. As Lenski (1966) has clearly documented, the end result is the creation of centers of power and privilege; and once extant, these centers seek to expand their power and privilege, causing centralization of power, increased inequality, and mobilization of coercive capacities to quell the resentments over such inequality of property and privilege.

Population pressures can operate via other economic paths to affect power. Jack Goldstone (1990), for example, documents the effects of population pressures on "state breakdown" in agrarian societies, as these increase resource scarcity, general price inflation, escalated costs to the state, increased state borrowing, raised taxes, state fiscal crisis, and opposition to the state from elites, while at the same time generating falling real wages (under price inflation), rural impoverishment, and mass protest movements. In figure 5.3 (Maryanski and Turner 1992:130), Goldstone's argument is summarized. Without necessarily accepting every detail of Goldstone's argument, which has generated controversy, the broad contours of his approach can nonetheless be generalized to all growing populations which begin to grow beyond the easy capacity of productive processes. When this happens, scarcity, price inflation, falling real wages, and state fiscal crisis ensure; and internal threats by both elites and masses inevitably increase. For elites feel their privileged status threatened (through loss of profits, wages, and patronage, coupled with increased tax assessments), whereas the masses sense that their subsistence is threatened (by falling wages, unemployment, and taxation). As the figure representing Goldstone's argument underscores, elite competition for patronage (or material incentives) from centers of power only serves to aggravate these processes, as does external war (which I have added to Goldstone's model to make it consistent with figure 5.2).

Thus, as I noted in chapter 3, population growth can not only

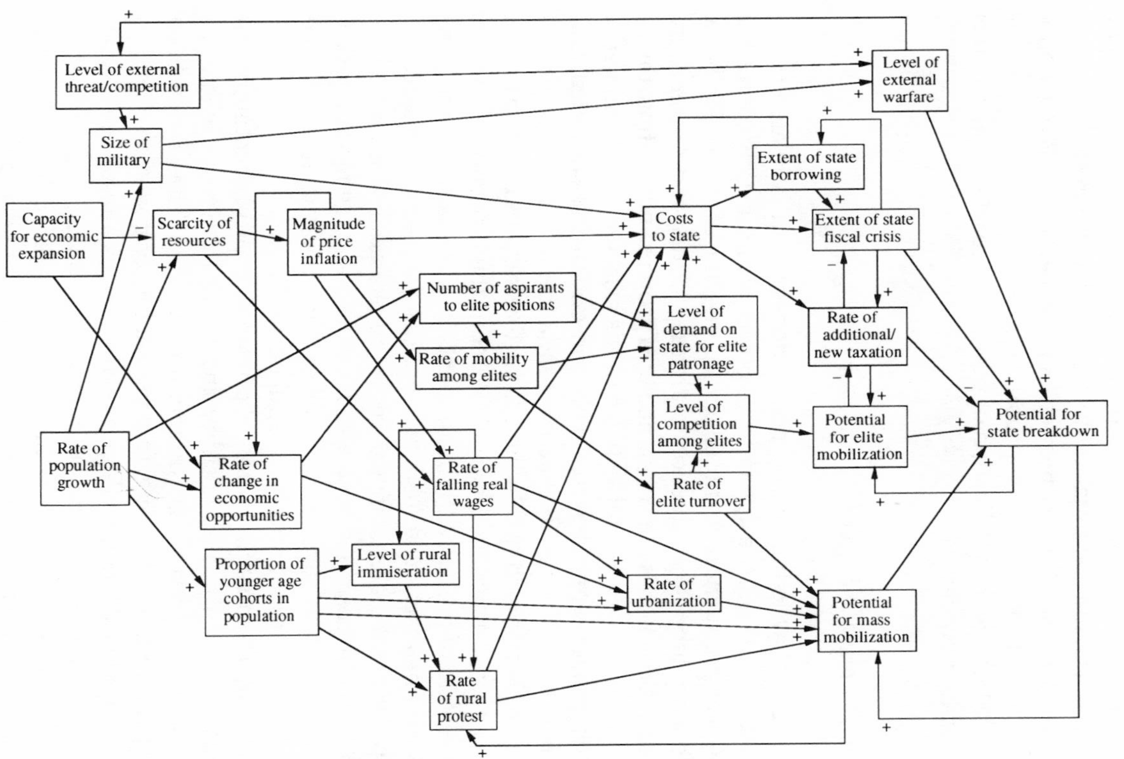

Figure 5.3. Goldstone's Model of State Breakdown

cause production and distribution to expand, but such growth often exceeds productive and distributive expansion, thereby setting into motion processes leading to de-consolidation of power, at least for a time. I will pursue these matters in more detail in chapter 8 on disintegrative processes, but for the present the general point is that population growth, as it affects production, is both a cause for the consolidation and centralization of power and for the breakdown of consolidated power.

Production and the Dynamics of Power

Regardless of whether or not population growth is related to expanded production, the dynamics of power are intimately connected to levels of production. As noted above, one line of connection is the relationship among production, material surplus, and power. If increased production is not consumed by a growing population, then a level of surplus beyond subsistence needs exists. Without a material surplus, power cannot be effectively consolidated or sustained, and so it is not surprising that the consolidation and centralization of power are very much tied to the productive surplus of a population (Lenski 1966; Turner 1972, 1984b). Without the existence of surplus and a system of property, the incumbents in administrative and coercive structures cannot be sustained, nor can the resources needed for manipulation of material incentives be generated. For as long as the incumbents in positions of power must also engage in purely productive activity to support themselves, the scale and scope of consolidation and use of power to regulate are limited.

The great dilemma for centers of power is how to extract sufficient surplus without arousing hostility and revolt from those who must give up this surplus. This "dilemma of taxation" has rarely been managed effectively for long, because centers of power also tend to be associated with the privilege of elites and, as a result, ever more surplus is extracted to support elite consumption rather than government, except perhaps for the coercive force needed to quell rising resentments. Agrarian societies were the most likely to get trapped into overtaxation leading to the "state-breakdown" outlined in figure 5.3. Modern political democracies have fared much better because of the dramatically escalated productivity and the resulting size of the material surplus that can sustain increasingly diverse activities, because of the growing amounts and diversity of property that can be distributed to all members of a population, and because of the increased responsiveness of elected incumbents in power to mass sentiments. Yet even here, taxation issues remain a source of tension between those in power and those who must pay.

Power does not become consolidated only because there is the surplus to pay for such consolidation and, eventually, centralization. There are heavy Spencerian selection pressures for regulation as production expands. One pressure is for a stable measure of value, so essential to the accumulation of productive capital. Thus, government must begin to issue and control money, and in so doing, it generates (*a*) a source of symbolic legitimation for itself (as long as inflation is managed), (*b*) a liquid source of revenue to sustain itself (as long as it does not overtax or overspend), and (*c*) a mechanism for accelerating the production and distribution on which its access to material resources depends. (I should add that the control of money also provides a common measure of property which can promote legitimacy if property is more widely distributed or, conversely, tension and instability if property is hoarded by elites. I might further add that democracy is a comparatively stable form of governance as long as it can sustain the sense, even if illusionary, that property is widely distributed.)

Another pressure for regulation comes from the entrepreneurial problems associated with expanding production. Once production moves outside of household units and employs a nonkin work force uses money and credit, and relies upon market distribution, the organizational dilemmas confronting a population escalate exponentially. Vast bodies of laws become necessary to regulate economic activity, courts must proliferate to adjudicate the escalating disputes among actors freed from traditional controls, and coercive force must be available to support laws and court decisions (Friedman 1959; Gurvitch 1953; Lloyd 1964; Turner 1980; Vaco 1988). Larger administrative structures become necessary to (*a*) regulate the arrays of new types of relations made possible with money, credit, and markets, (*b*) manipulate material incentives, and (*c*) collect and distribute greater amounts of tax revenues; and again, coercive force must be ready to enforce administrative decisions.

Still another pressure for regulation comes from the inequalities and conflicts created by expanded production and consolidation/centralization of power. As Lenski (1966) documents, inequality and stratification increase with rising production, as elites use power to horde resources; and it is only with conflicts over resources in advanced agrarian systems that political democratization ensued. The potential for conflict, or what I labeled "internal threat" in figure 5.2, is most likely to cause increased mobilization of coercive capacities, although expansion of administrative structures also occurs with efforts at controlling internal threats.

Yet another pressure for regulation stemming from conflicts associated with increased production comes from the differentiation of larger numbers of productive organizational units whose interests and goals often collide. Such conflicts of interests inevitably force the consolidation of administrative and coercive structures; and they frequently initiate efforts to mobilize and use symbols and incentives to control these conflicts. For if conflicts become too intense, they pose an intense internal threat which can destroy efforts to consolidate power and, thereby, disintegrate the population. Conflicts of interest thus work to reduce the capacity to consolidate power. As such conflicts escalate and become perpetual in highly differentiated populations, regulation and control become problematic and often reveal oscillations between efforts at tight administrative/coercive solutions followed by decentralization of control when tight control proves problematic and potentially delegitimating. Modern political democracies in high volume/velocity market systems are particularly vulnerable to these oscillations because such markets generate large numbers of conflicting interests and other logistical problems but, at the same time, place demands (both ideological or symbolic and politically) for less regulation.

Distribution and Power

Distributive dynamics are intimately connected to those affecting power—as I noted in the closing pages of chapter 4. As the level of distribution among actors in a population increases, the relative importance of the four bases of power changes. For when the distributive infrastructure expands and as exchange distribution relies on money and market transactions, the scale of political organization can dramatically increase. As Weber ([1922] 1978) emphasized, the existence of money provides a liquid source of wealth which can be used to sustain administrative structures (i.e., pay bureaucrats) and coercive capacities (i.e., pay and support soldiers, police), while potentially providing resources for manipulation of material incentives and symbols. Moreover, by controlling the printing and distribution of money, a much more flexible and far-reaching capacity for using material incentives emerges, and a new set of symbols—money as a generalized medium of political legitimation—becomes available.

Market exchanges accelerate these changes because money, property, and capital can be rapidly accumulated and then taxed as a basis for supporting the extension of power-use; and because money becomes so pervasively associated with value (in exchanges), it increasingly supplants other, more traditional symbols previously used by centers of power to control, regulate, and legitimate.

But markets and money are a double-edged sword: markets can cycle into dramatic downturns and are potential arenas of fraud, corruption, and overspeculation; whereas money is subject to inflation, while accelerating the dangers posed by markets. As a consequence, the wealth and resources on which power depends for financing its administration, its capacity for coercion, and its ability to manipulate incentives can rapidly disappear with market downturns, or even collapse, while a major resource for symbolic manipulation and political legitimacy can quickly erode (in contrast to more traditional and stable symbols, such as those symbols associated with religion). Thus, although money and markets enable the scale of power-use to expand, the potential volatility of these distributive forces also poses hazards for centers of power.

Moreover, because markets are a principal force behind sociocultural differentiation, they create not only selection pressures for coordination and control among an increased diversity of units, but they also generate potential centers of counterpower. For new units, resource niches, and alliances of units in and between niches can all become a base for political mobilization, and hence, they can reduce the level of consolidated and centralized power regulating a population (Rueschemeyer 1977). Indeed, political democracies are most viable where market systems are developed because differentiated markets constantly produce interest-group politics and liberal symbols that work to prevent power from consolidating too much administrative, coercive, and symbolic control, while sustaining a system of material incentives outside the bases of political governance in a society.

Money and markets also shift the relative weights of the four bases of power. With money to pay administrators this base of power increases in importance—as Weber ([1922] 1978) stressed in his discussion of "rational-legal authority." Money and markets not only enable and facilitate this process of administrative expansion, they generate Spencerian selection pressures for such expansion. Market differentiation and potential for downturns, fraud and corruption, elaboration of credit and financial instruments, expansion of contracts, and escalation of disputes all create problems of coordination, control, and regulation which, eventually, can only be met by the creation of new administrative, adjudicative, and enforcement structures. And yet, as noted above, markets also generate sources of counterpower that prevent overregulation through too much centralization of power.

With a liquid resource like money, the importance of manipulation of material incentives by centers of power also increases. Threats of increased taxation on property and offerings of tax incentives or punishments can now be effectively employed to regulate and control activities. Similarly, dynamic and differentiated markets provide many alternative arenas for manipulation of material incentives through not only tax policies but also spending and borrowing patterns by centers of power in selected markets as well as by regulatory actions (or threats of such actions) in various market niches. Again, market dynamics and differentiation not only provide the opportunities for new and expanded manipulation of material incentives, these forces require such manipulation by virtue of the need to print money and the necessity to regulate borrowing, financing, contract disputes, fraud, abuse, exploitation, and periodic crises in market transactions.

The use of money and markets increases the importance of administrative bases of power and the ever-growing reliance on material incentives by virtue of creating demands (a form of Spencerian selection) for infrastructural development to facilitate the use of money and the expansion of exchange transactions. The building and upgrading of transportation and communication systems, plus an array of regulatory agencies to monitor the flow of money and the volume of transactions, all cause the administrative component organizing centers of power to grow, while providing new arenas for manipulating incentives through such practices as letting of construction contracts, purchase of particular goods and services, or borrowing through choices among various instruments of credit to finance infrastructural development.

Thus, as the level of distributive activity—both infrastructural and transactional—grows, there are selection pressures for the administrative and incentive bases of power to become more significant than either the symbolic or coercive. And, if the reverse is true—the symbolic and coercive dominate—then distributive activity, especially exchange dynamics, will be thwarted. Particular configurations of coercion and symbolism, as these are linked to the administrative base of power, produce very different patterns of power-use. If the coercive and administrative bases of power are linked, then a totalitarian pattern of power-use is evident. If traditional and typically religious symbols are tied to the administrative, then a theocracy, like modern-day Iran, exists. Both of these profiles, however, are unstable if exchange processes are well developed, because the existence of money in differentiated and dynamic markets represents a source of resistance to overregulation and nonsecular orientations. To sustain a totalitarian or theocratic level of regulation inevitably leads to productive and distributive stagnation which, in the long run, will weaken a political regime and eventually encourage internal conflict or invasion by a more productive population. This is why the relationship delineated in figure 5.2 between centralization of power and distributive processes is positively curvilinear; a certain amount of centralization is essential for effective regulation, and indeed unregulated markets generate crises and collapse that inevitably cause centralization of power-use, but overregulation leads to stagnation of both exchange and production—as Herbert Spencer, Gaetano Mosca, Georges Sorel, and Vilfredo Pareto clearly recognized long ago. In fact, when exchange processes reveal high levels of differentiation and a high velocity/volume of transactions, they operate as a source of resistance to excessive centralization of power and, potentially, as centers of counterpower.

Inequality, Stratification, and Power

Inequality is the term I use to denote the degree to which the distribution curve for each valued resource among members of a population is skewed. As such, the analysis of inequality involves constructing a separate distribution for each valued resource, such as material capital, power, prestige, or anything that is highly valued by the members of a population. The existence of a system of property—defining what is to be possessed and by whom—is essential for inequalities to reach high levels; and so, it is with agrarian systems, in which property becomes more clearly defined and diverse, that inequalities begin to escalate dramatically.

Stratification involves the creation of subpopulations, or "classes," which are distinctive in their shares of resources, especially property. Stratification shifts attention from distribution to additional questions: (*a*) how correlated are these separate distributions of resources (i.e., to what degree are those high or low in one distribution also high or low in other distributions) and (*b*) how much mobility up, down, and across distributions occurs. When there is great discontinuity in the share of resources, when there is a high level of correlation across distributions among those receiving particular shares, and when mobility rates are low, inequality is high and stratification is likely to be rigid; and conversely, when there is less correlation across distributions and high rates of mobility up, down, and across distribution is high, inequality is lowered and stratification will be less rigid (Turner 1984b). Inequality and stratification will be examined again in chapter 8 on disintegrative forces, and so, my focus here is narrow. How do high or low inequality and rigid stratification interact with power?

The general view of most sociologists might be termed the "usurpation model": those with power extract resources from the less powerful, and in so doing, they increase inequality. Such usurpation requires, of course, a material surplus to usurp, and so, the effects of power on inequality are very much tied to the level of production and distribution as these create material surplus and systems of property. As Lenski (1966) argued, such a usurpation scenario had been the evolutionary story of human populations since hunting and gathering up to advanced agrarianism, with a downturn under industrialization. Both the consolidation and centralization of power have the effect of increasing inequality. As the administrative, coercive, and incentive components of power are consolidated, resources need to be extracted; and as centralization of the administrative structure occurs, further extraction occurs to support larger chains of command and larger agencies of regulation. At the same time, elites in a system of consolidated and centralized power are able to secure some portion of this extraction, via a variety of processes—for example, exerting influence on the decisions of proximate leaders, securing favorable franchises and concessions, operating as designated tax collectors, or threatening revolt and opposition.

One of the persistent dilemmas for elites in the process of centralizing power is that they must create incentives for skilled and professional administrators to carry out their dictates. Otherwise, and this has been the more frequent case historically, resources that could go to elites are "lost" through inefficiency, incompetence, and corruption by those collecting taxes, operating franchises, enforcing decisions, and performing other necessary administrative functions. Historically, a variety of strategies have been used by elites to secure competent administrators—from scholarly mandarins through castrated scribes to examination-graded civil servants—but if elites must use material incentives to attract competent administrators, the level of inequality can be reduced by the portion of resources necessary to sustain administrative competence. Moreover, once these administrators are in place, and elites become dependent upon them, they can demand more resources, thereby further reducing the privilege of elites. Yet without a broader democratization of the political process as a whole, this system of administrators can demand only so much before elites cease to make further material concessions. Thus, inequality will remain high; and in fact, if administrators are effective, they can extract ever more surplus resources for elites and thereby increase inequality.

As this process of concentrating resources occurs, the degree of inequality increases and stratification becomes more rigid, reaching its most extreme proportions in advanced agrarian populations where a hereditary nobility usurps most resources from a large peasant class and controls virtually all property. Production and distribution, however, can mitigate against such extreme inequality, to the degree that they cause differentiation of structural units and niches in which these units can secure resources. Moreover, by creating niches, intervening "classes" between the elites and masses emerge, thereby decreasing the polarization of the population into "haves" and "have-nots." And as noted above, a cadre of administrators for elites adds yet another intervening class. The effects of this differentiation, as fueled by production and market distribution as well as expansion of administrative functions, become evident with the transition to industrialization, thereby working to reduce somewhat levels of inequality.

A more volatile force is the internal threats produced by very high levels of inequality and stratification, as subordinate subpopulations periodically revolt, while posing a constant source of potential conflict. Ironically, revolts or potential for conflict cause, at least initially, increased centralization of power and mobilization of coercive capacities, which, in turn, require more usurpation of resources to be sustained. Yet as ever more inequality is generated, even elites may pose internal threats if their resource shares are taxed; and added to the growing mobilization of coercive powers among the masses, coupled with the withdrawal of political legitimacy, internal threats erode both the consolidation and centralization of power. Additionally, if productive and distributive units become overregulated and overtaxed by centers of power trying to deal with internal threats, these too become a potential source of internal threat and counterpower.

It is the combination of (*a*) differentiation of niches and structural units by growth in production and expansion of distribution and (*b*) rising internal threats that account for the decrease in inequality with industrialization. A diversity of structural units seeking resources mitigates against extreme polarization in resource distribution, and a persistent source of internal threat that can potentially destroy centers of power or, at the very least, undermine legitimacy eventually causes redistribution of some resources. When these pressures fall short of revolution and civil war (which generally serve to produce new centers of coercive power) and generate some form of political democracy, then the level of internal threat is dramatically reduced because there is a political mechanism for preventing extreme inequality and for challenging excessive consolidation and centralization of power. And, with the corresponding relaxation of tight regulatory control on production and distribution (including the production and distribution of ideas via media, educational structures, and voluntary associations), many new resource niches cutting across diverse resource distributions and creating new, amorphous, and overlapping layers of stratification operate to reduce internal threat (Lenski 1966; Parsons 1966, 1971).

Geopolitics and Power

Just as internal threats and conflict cause power to become centralized, especially its administrative and coercive bases, so external threat generates pressures to centralize administrative functions and mobilize coercive forces to deal with the threat (Spencer [1874–96] 1898; Carneiro 1970; Webster 1975; Fried 1967). Such shift in the bases of power to a coercive-administrative profile is also accompanied by efforts to intensify collective symbols to legitimate mobilization for conflict. Thus, as a population shifts the bases of power to deal with external threats, several processes are set into motion.

As we will explore in more detail in the next chapter on spatial dynamics, the level of coercive mobilization and the degree of success in external conflict are related to the degree to which, relative to adversaries, a population is sufficiently large, productive, and politically organized to fight successfully. And, if centers of power can win wars, their level of political legitimacy is increased, thereby enabling further consolidation and centralization of power. Yet such consolidation and centralization will, as Spencer ([1874–96] 1898) first argued theoretically, eventually cause productive and distributive stagnation, while increasing inequality and internal threats. Such threats can begin to erode legitimacy and the supply of resources necessary to sustain military mobilization—as leaders in most feudal societies were to discover and as, more recently, leaders of the old Soviet Union learned. Loss of a war, as Weber ([1922] 1978) and later Skocpol (1979) argued, accelerates these delegitimating processes, reducing the symbolic base of consolidated power and unleashing internal threats into open resistance to coercive and administrative centers of power. The loss of a war is, eventually, inevitable for as rates of external conflict increase and new territories are acquired, logistical loads for transportation, communication, and administration increase; and as conflict extends a population's boundaries, this population will eventually encounter an enemy, or coalition of enemies, that can cause a loss or at least military stalemate, either of which erode legitimacy and unleash the accumulating internal threats (Collins 1986). Moreover, if a society has been successful in some wars and has, as a result, conquered and annexed diverse populations, the level of internal threat has been rising with each new conquest, thus further setting the stage for political collapse (although in the short run such threats will cause further centralization of power and mobilization of coercive capacities to deal with the internal threats).

Both external and internal threats increase logistical loads, setting into motion Spencerian selection for more production, distribution, and consolidation of power. Yet as power is consolidated and centralized around its administrative-coercive bases, it sets the conditions—productive and distributive stagnation, overregulation, increased inequality, escalated internal threats, and geopolitical overextension—that eventually cause the de-consolidation of power. There is, then, an inherent dialectic in the centralization of power.

The Dialectics of Power

Without the consolidation and centralization of power, the organization of larger populations over time and space would be impossible. Those populations under Spencerian selection that could not develop and sustain consolidated power disintegrated, or they were conquered by other populations which more effectively consolidated power. It is for this reason that Herbert Spencer ([1874–96] 1898) contended that war had been a driving force of societal evolution: the more organized populations selected out the less organized, or incorporated them into more complex structural forms.

Yet the consolidation of power is usually precarious, at least in the long run. Achieving a balance among the bases of power—administrative, coercive, symbolic, and incentive—is always difficult. Too much in the way of coercion, administration, incentives, or symbolic intensity creates distortions which make political regimes vulnerable. Moreover, because power is about control and regulation, its use is always resented by some portion of a population, and the overuse of any base of power will set off resistance by segments of a population.

Thus, the history of all large-scale populations reveals the consolidation and centralization of power, followed eventually by de-consolidation of power. During this cycle, the relative salience and reliance on different bases may shift, but in the long run in all human populations de-consolidation of power, often accompanied by actual disintegration of the population, has occurred. There are, of course, counterarguments about "the end of history" as the world's societies gradually move toward liberal political democracies, but these arguments assume that the "perfect" balance among the bases of power has been achieved, that internal and external threats have now been managed, and that cycles of production and distribution have been mitigated to the degree that the dialectics of power have been suspended. As effective as Western democracies have been in achieving and sustaining a balance among the bases of power during periods of moderate centralization of power, there is no reason to assume that such systems are the end of history any more than headman systems, chiefdoms, and feudal systems—all of which had longer histories than political democracies—were the end point of the past. Moreover, there are sound theoretical reasons for being skeptical that the perfect balance of power has been achieved by the Western democracies.

First, power is always resented, and its use does not eliminate the inequalities that fuel internal threats. Second, overpopulation will continue to pose intense logistical loads on many populations in the world, which will set into motion Spencerian selection for highly centralized and coercive centers of power to regulate the inequalities and material shortages that come with too many people placing burdens on production and distribution through their demands for consumption of surplus capital. Third, geopolitical conflict will continue to distort political regimes toward coercive-administrative profiles, thereby making them vulnerable to potentially rapid delegitimation with the loss of wars or with declines in domestic production and distribution as these create material shortages and aggravate inequalities. Thus, it is unlikely that there is an end to history; power will continue to be consolidated and centralized, only to collapse and then be reformed again.

Elementary Principles on the Dynamics of Power

Power is the capacity to regulate and control members of a population, as well as the units organizing these members' activities. At the macro level, this capacity involves the consolidation into varying configurations of the four bases of power—manipulating cultural symbols, bestowing material incentives, using coercive force, and erecting administrative structures. Consolidation initiates the centralization of power, although other forces are also critical in affecting the degree of centralization as well as the profile of consolidation.

The basic dynamics of power can be summarized as a series of elementary propositions:

I. The degree to which the four bases of power become consolidated into patterns of political leadership is a positive and additive function of:

- A. the level of Spencerian selection pressures stemming from the level of logistical loads, which are a positive and additive function of:
    1. the size of a population
    2. the level of external threat
    3. the level of internal threat

4. the level of entrepreneurial problems in production and distribution
5. the level of structural differentiation among the units organizing members of a population

B. the level of material surplus available, which is a positive and multiplicative function of:

1. the level of production
2. the level of distribution

C. the degree of centralization of power, which is a:

1. positive curvilinear function of the level of internal threat
2. a positive function of external threat
3. a positive function of the degree of coercive mobilization
4. a positive curvilinear function of the level of distribution
5. a lagged positive function of the level of previous consolidation of the four bases of power

II. The stability of consolidated and centralized power and, hence, the capacity of centers of power to regulate effectively is:

A. a positive function of the balance among the four bases of power, which, in turn, is a negative function of the degree to which any one base (coercion, incentives, symbols, or administration) is disproportionately used to regulate, especially with respect to symbols or coercion

B. a lagged negative function of the degree of regulation by centers of power of production and distribution processes

C. a negative function of the level of inequality and resulting increases in internal threat

D. a positive function of the degree and persistence of external threat and the continued success in external conflict, up to the point where principles II. A. and II. B. are violated, which then transform the relationship to negative

6

SPATIAL DYNAMICS

SPATIAL DYNAMICS REVOLVE around a number of interrelated phenomena: the distribution of a population in a territory, the density of such distribution, the movement of individuals within and between territories, and the control of territory. When conceptualized in this way, two rather separate lines of theorizing become relevant: urban ecology and geopolitics. The former can provide insights into the distribution of a population within a territory, whereas the latter helps us understand the amount of space and territory controlled by a population.

In presenting a separate analysis of spatial dynamics, I want to emphasize that the amount of territory controlled by a population and the movement and distribution of people within this territory exert independent effects on other macro-level forces. True, spatial dynamics flow out of other macrodynamics, but they have important reverse causal effects on these dynamics. In so doing, they exert effects on the organization of a population that need to be conceptualized not as epiphenomena of other forces but as forces in their own right.

Urban Ecological Theorizing on Spatial Dynamics

Figure 6.1 presents a composite of these various meso-level theories in "urban sociology." This model does not conceptualize the dynamics of urban areas in the same detail as much modern urban ecology (see, for example, Kasarda 1972; Hawley [1971] 1981; Frisbie 1980; Berry and Kasarda 1977). Rather, I focus on (1) the concentration of the population (i.e., the density of people with spatially bounded areas), (2) the level and rate of geographical expansion (i.e., the movement of individuals outward from concentrated core areas), and (3) the overall level of agglomeration (i.e., the extent to which the population as a whole is located within densely settled spatial areas, many of which are contiguous and connected). Much urban sociology, and urban ecology in particular, specifies the dynamics of these processes in fine-grained detail, but my goal in presenting the model is to use the ideas of urban ecology to inform a more general theory about the organization of varying numbers of individuals in varying amounts of space.[1]

Following Émile Durkheim's lead, one set of key variables in most modern ecological models is communication and transportation technologies, which are used to develop what I termed the "distributive infrastructure" in chapter 4. As these technologies develop, they are used to expand the material infrastructure for transportation (roadways, canals, ports, railroads, airports, etc.) and communication (which historically overlap with those for transportation, except in recent history with the advent of the material means for the "information age"). The density of a population is, to some degree, related to the capacity to move people and information, which is connected to the ability or willingness of productive units to generate a material infrastructure revolving around not only transportation and communication systems but also physical facilities for all types of activity—housing, production, exchange, governance, arts, religion, and other activities that require physical structures and infrastructures.

The expansion of production and distribution generates, as noted in chapters 3, 4, and 5, not only the wealth to sustain separate centers of power but also new types of second-order logistical loads. Many of these come as production and distribution reach high levels, but in the end further consolidation of power is required. As power is consolidated and centralized, it too can create second-order logistical loads. And as the scale of the material infrastructure expands in response to production, distribution, and power, additional second-

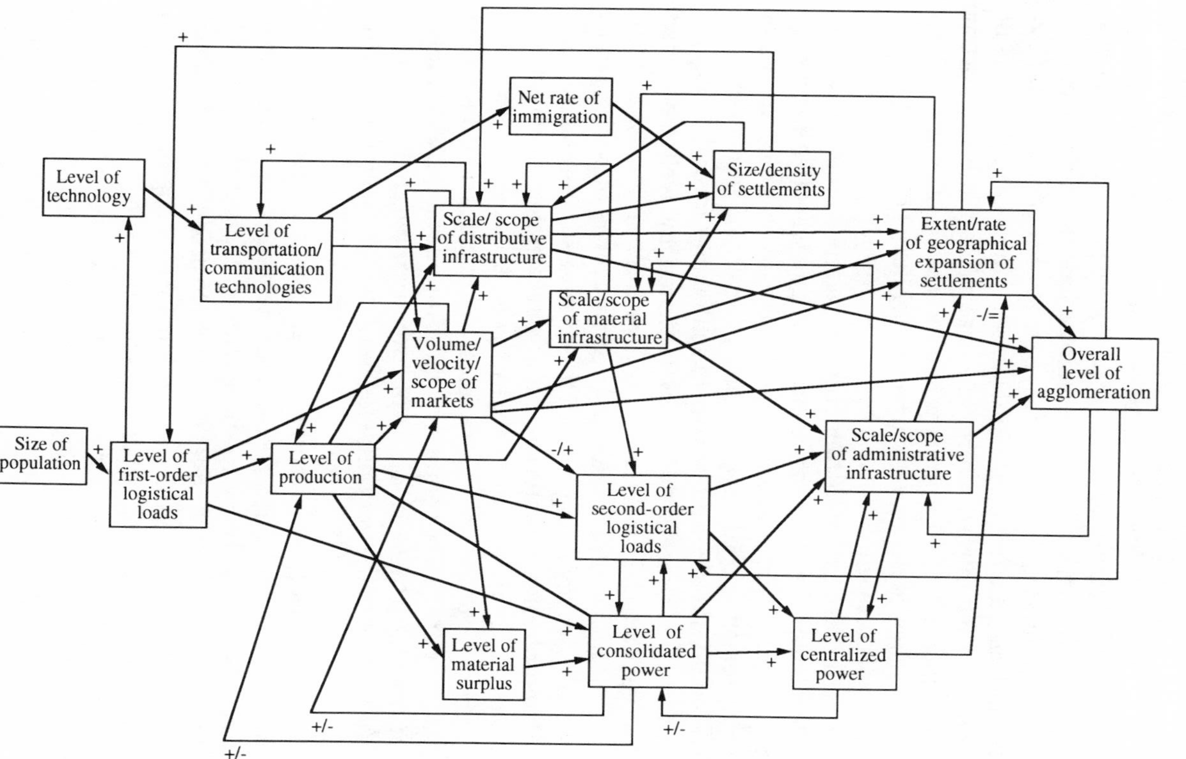

Figure 6.1. The Abstracted Urban Ecology Model

order logistical loads are created. When power is consolidated and centralized, it expands the administrative infrastructure revolving around decision making, adjudication, and control. Moreover, as increases in productive and distributive forces begin to impose new second-order logistical levels, nongovernmental administrative structures expand (e.g., administrative structures in production, exchange, and distribution operations), further increasing the scale of the administrative infrastructure. The expansion of the administrative infrastructure is reciprocally related to the scale of material infrastructure, being dependent upon it while, at the same time, stimulating its growth.

The size and density of settlements are positively related to these forces. Moreover, the distributive infrastructure will often encourage high rates of immigration, which, generally, involve movements of people into existing settlements, thereby increasing settlement density. Eventually, the size and density of settlements force their expansion outward, a process which is facilitated by well developed distributive, material, and administrative infrastructures. Initially, this outward movement is often resisted by centers of power, but as market processes revolving around the relative supply and demand for land and physical facilities eventually accelerate movement outward. Indeed, market forces will, in the end, outweigh efforts by centers of power to control geographical expansion; and in fact, decision makers will often be coopted by these forces and facilitate movement outward. All these processes of "urban sprawl" reduce density, but only relative to the very high levels of density seen during early urbanization.

These processes have important reverse causal effects on other macrodynamic forces. First, dense settlements generate Spencerian selection pressures, as Boserup argued, for an expanded distributive infrastructure to move resources from rural to urban areas and, eventually, for a more extensive market system for exchanging goods and services within expanding urban areas and between both urban and rural areas. And these forces exert mutually positive effects on each other as increased market activity encourages expansion of the distributive infrastructure, and vice versa. Second, these distributive processes create additional wealth which can be usurped by centers of power, thereby increasing the consolidation of power around its administrative base. Third, dense settlements also create new kinds of social control problems stemming from several potential sources of counterpower: (*a*) the concentration of potentially restive and readily mobilized urban masses, (*b*) the emergence of a wealthy elite engaged in exchange distribution that could challenge traditional authority, and (*c*) the old elite, who are increasingly threatened by (*a*) and (*b*) plus the growing concentration of administrative and coercive power to regulate threats from sources of counterpower in dense settlements. Thus, density of settlements generates logistical loads revolving around control problems which, in turn, magnify Spencerian selection for increased concentrations of coercive/administrative power and which, ironically, further consolidate and centralize power in ways that increase density of settlements.

The particulars of urban sociology[2] seek to explain the development and the precise form of settlement for a highly agglomerated population (e.g., Zipf 1949; Clark 1951; Hawley [1971] 1981; Berry and Kasarda 1977:95–97; Stephan 1979a, 1979b; Klassen, Molle, and Paelinck 1981; Hall 1984, 1988; Castells 1985; Frisbie and Kasarda 1988). For my more macro purposes, however, the key variable is the general level of agglomeration, with this level being a joint function of the variables enumerated in the model portrayed in figure 6.1: density of population in spatial areas, scale of material, distributive, and administrative infrastructures, and volume and velocity of markets. What I argue, then, is that increased production, growth of the material and administrative infrastructures, and expansion of markets cause increased density of settlement and, over time, geographical expansion in the form of contiguous urban settlements *and* larger numbers of smaller, noncontiguous, but still dense settlements. Very high levels of communication and transportation technologies as these affect the distributive and material infrastructures as well as the velocity of markets accelerate geographical expansion as well as the development of larger numbers of noncontiguous areas of dense settlement which are, nonetheless, integrated into larger and more densely settled urban cores (this latter process can extend to the world-system level, as is evident today, where different populations are, to a degree, agglomerated by virtue of their functional interdependence with large urban centers in different parts of the world).

I believe that these rough generalizations will apply to more than the "modern world" and can account for processes in less-developed regions in the present as well as in the distant past. That is, populations of the past or present with low productivity, little material or administrative infrastructure, low levels of consolidated power, and low volume/velocity markets (the situation in most parts of the world until relatively recent history) will be (*a*) smaller and less densely settled in space, (*b*) less likely to expand geographical space outward from existing urban cores (assuming they have them), and (*c*) less agglomerated into a variety of dense urban centers. Populations with moderate levels of productivity, some material and administrative infrastructures, some degree of consolidated and concentrated power, and viable markets revealing some volume and velocity (and this is the case for most of the "nonmodern world" today and most of the preindustrial world three hundred years ago) will evidence dramatic increase in the size and density of a few urban cores, some degree of expansion from these cores, and potentially, an increase in the number of noncontiguous urban centers, and only relatively low levels of agglomeration.

Although these kinds of generalizations may be time-bound, especially as it is difficult to know what the extremely high loadings for population size, production, markets, and infrastructures will bring about in the future, the model in figure 6.1 argues that, despite local and historically unique variations in the precise patterns, increase in the size, density, extension of urban space, and agglomeration will ensue. And such agglomeration has reverse causal effects—some of which are direct, others indirect—on population, production, distribution, and power dynamics.

Geopolitical Theorizing on Spatial Dynamics

From a macro perspective, spatial dynamics involve more than the distribution of a population in settlements; they also involve interpopulation relations influencing the cultural/ethnic diversity of a population, its size, and its expanse of territory. Thus, to complete a macro structural picture of spatial processes, it is necessary to examine geopolitics revolving around conflict and conquest of one population by another, for out of such conflicts come varying patterns in the size of territories controlled by a population. Among the various theories that could be examined Randall Collins's (1986) analysis of geopolitics is the most relevant to a macro structural theory of space.

Much like Spencer's early theorizing on "militant societies," Collins (1986:167–212) views internal system processes and external geopolitical activities as interconnected. Geopolitical empires are created by war, but success in war depends upon the level of productivity, population size, resource levels, capacity to mobilize and legitimate concentrations of power, and most significantly, extent of "marchland advantage" (i.e., the degree to which a population is "protected" from "enemies" by natural barriers and, as a result, can fight wars on only one front, or on a small proportion of its borders).

In figure 6.2, the dynamic relations among the variables in Collins's theory are modeled. High levels of coercive power can only be mobilized by high productivity and surplus wealth, which are related to levels of technology, resource advantages, and population size. The mobilization of coercive power can be sustained by success in conflict, which, as Weber ([1922] 1978) emphasized, legitimates the use of power and which, at times, provides additional resources and wealth (through plunder and exploitation). But, much as Spencer recognized, success in warfare extends the territorial space of an empire, which, eventually, works against further expansion of territory. For as the extent of territory increases, several countervailing forces are set into motion: (1) the level of ethnic diversity increases (as conquered peoples are added to the population) and creates problems of potential revolt, which increase the logistical loads revolving around control, co-optation, and coordination of diverse and restive subpopulations; (2) the level of logistical loads also increases as a result of having to move materials (supplies, weapons), people (administrators, soldiers), and information (orders, directives, guidelines) across larger territories, farther and farther removed from centers of power and administration; (3) the marchland advantage is eventually lost as an empire expands and must confront enemies on more than one front and, inevitably, another empire (creating the possibility of a showdown war); and (4) the technological advantage will also be lost as enemies copy those technologies (both economic and military) that enabled one population to subjugate and control another.

Under these escalating conditions, territorial expansion reaches the point of overextension, as logistical loads increase and previous marchland and technological advantages are lost. And, even if overextension is not a problem, the likelihood of a showdown war with another empire or confederation of threatened societies increases with territorial expansion. Thus, there is a built-in corrective to geopolitical processes across space: at some point, it becomes difficult to conquer and control additional populations and territories. For this basic reason, Collins argues that empires have historically reached a maximum size of about four to six million square miles and, then, begun to disintegrate as their point of overextension was reached or as they became involved in a showdown war. This figure for the maximal size of empires may be too low, for the largest empire known—that created by the Mongols—was considerably larger than six million square miles. But the general point is correct: at some point empires reach their point of overextension, as was the case for the Mongols when they sought to conquer Japan, or as was true for the Soviet Union when it engaged in an arms race with the more-productive West.

The reverse causal chains in the model can also indicate why empires often collapse or begin to implode back on themselves at the point of overextension. Once there is little chance of success in external wars, several chains of events unfold: The capacity to hold territory and control populations decreases; the ability to extract wealth to finance power and its mobilization into coercive force declines; the legitimacy of centers of power is eroded as success in the geopolitical sphere declines, setting the stage for coups or revolts; and given that mobilization of coercive capacities skews production to military needs, while discouraging technological innovation in the domestic sphere and usurping capital from the domestic economy, the economy becomes incapable of expanding production and generating increases in the economic surplus and wealth upon which coercion and its legitimation depend. The result is that the empire disassembles slowly, or if it loses a showdown war, the process is accelerated. Moreover, I might add that once collapse of a geopolitical empire has occurred, it is virtually impossible for it to reemerge, primarily be-

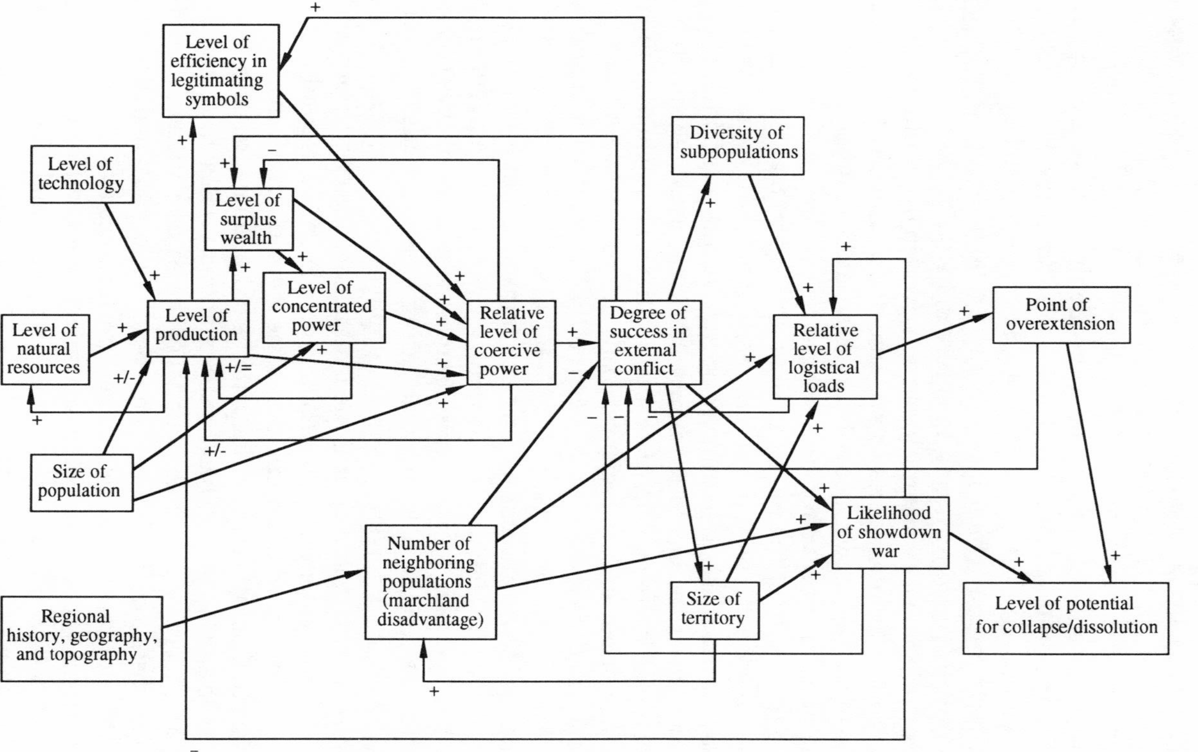

Figure 6.2. Collins's Theory of Geopolitics

cause the forces that produced it—technological advantage, high production, legitimated power, resource and marchland advantages, and coercive capacities—have been reduced and, in the case of showdown war, destroyed and plundered.

There is, then, a kind of rhythm to geopolitical processes: populations expand their territory and collapse at the point of overextension, only to have the geopolitical vacuum reorganized by another expanding empire, which, inevitably, will collapse. Because empires have difficulty reconstructing themselves, I would add to Collins's theory the hypothesis that the long-term trend in the world has been for large territorial empires to stagnate and decline, with warfare increasingly fought over geopolitical boundaries among smaller territorial units in various regional arenas.

The dynamics of the model presented in figure 6.2 also help explain specific patterns and configurations of territorial organization. Those empires that had to traverse oceans and other significant ecological barriers, or in the modern world, that must rely upon air technologies to cover long distances, are the least stable over time, primarily because their logistical loads are so high. Movement of materials and people becomes very costly the greater the distance from a home base, especially if natural barriers as well as distances require heavy reliance of air and sea transportation technologies. Moreover, while high levels of technology facilitate the movement of materials and people over long distances, the costs of these technologies are very high, thereby increasing logistical loads even further. The organization and control of populations in large territories are thus considerably easier if the land masses are contiguous and allow the movement of materials and people on the ground and if there are few natural barriers to escalate logistical problems.

Thus, in Collins's theory, the patterns of building territories tend to revolve around several basic considerations: (1) Resource, marchland, and coercive power advantages of some populations over others; (2) distance from the home base, or capital region, with increasing distance placing an ever greater logistical burden on a population; (3) points of potential contact between two expanding empires; and (4) previous patterns of integration and disintegration of populations that once engaged in territorial expansion. And so, although there is certainly much that is historically unique to geopolitical processes, there are basic underlying processes which are subject to theorizing and which can, therefore, become part of a general theory of spatial dynamics.

Geopolitical processes exert important effects on internal spatial patterns within a territory along two general sets of causal paths: (1) those revolving around the effects of geopolitics on the centralization of power as this works directly and indirectly to determine the scale and scope of the distributive, material, and administrative infrastructures, and (2) those revolving around the effects of geopolitics on conflict and its effects on the migration of people within a territory. We begin to see these and other effects by synthesizing geopolitical and urban ecology theories.

A General Model of Spatial Dynamics

Figure 6.3 pulls together the arguments from theories on urban ecology and geopolitics. The model seeks to link spatial dynamics to other forces examined thus far, namely, population, production, distribution, and power. Each of these forces exerts direct and indirect causal effects on spatial arrangements, but equally significant, the patterns of settlement within a territory and the extent of the territory exert important reverse causal effects on other macrodynamic forces. Because this is one of two highly complex models presented in this theoretical effort, let me repeat what I emphasized in chapter 1 about the difference between description and explanation. The model is complex not so much because I am describing a particular empirical case but, rather, because I am showing how many generic forces converge in explaining spatial processes. Obviously, the model is too complex to test as a whole, but particular causal paths can be reformulated in more testable propositions.

The geopolitical portion of spatial dynamics is portrayed along the lower half of the model, whereas the abstracted urban ecology portion is arrayed across the middle and top sections of the model. The two meet via the direct, indirect, and reverse causal arrows connecting the lower to the middle and upper parts of the model. Geopolit-

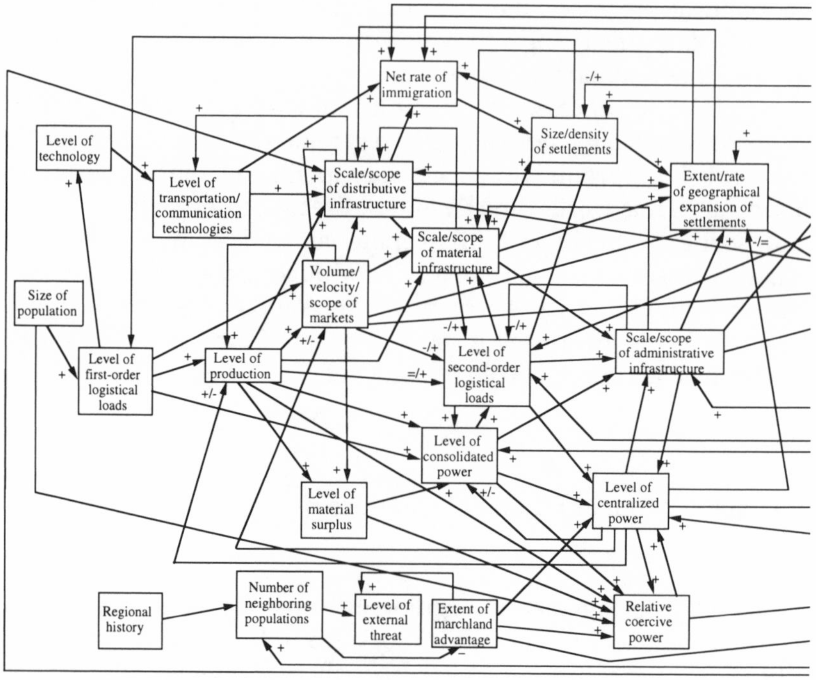

Figure 6.3. A General Model of Spatial Dynamics

ical activity sets a number of forces into motion that, in turn, influence the settlement patterns within a territory. Let me explore the most important of these.

First, as power becomes centralized as a consequence of external and internal threats as well as first- and second-order logistical loads, the administrative infrastructure is dramatically expanded. Such expansion creates needs and opportunities for individuals in the administrative structure of government, thereby encouraging immigration and larger, more dense settlements where the functions of governance are executed. In turn, this expanding administrative system directly expands the scale and scope of the material infrastructure, which encourages migration into, while providing the physical facilities for, larger and more dense settlements. Moreover,

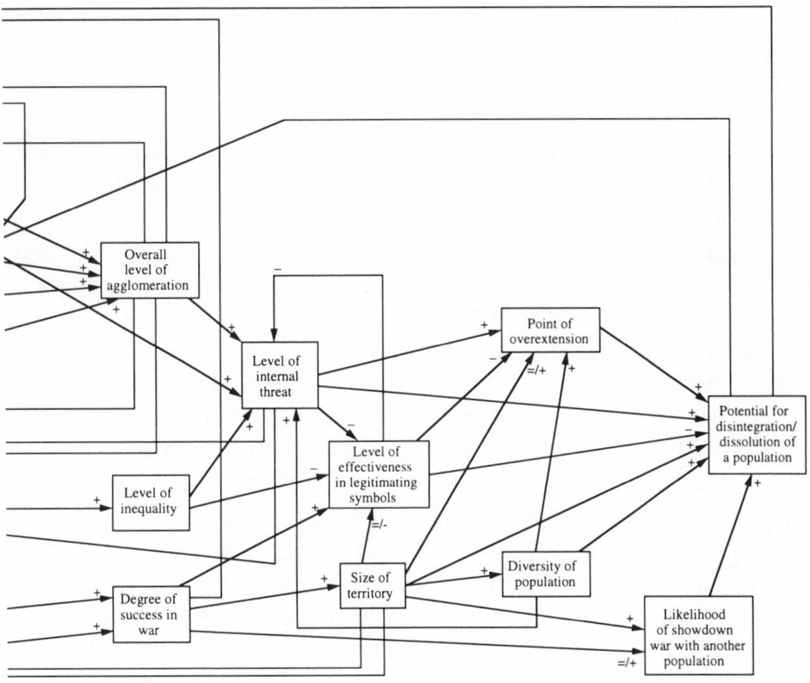

with more extensive administrative and material infrastructures, the geographical expansion of settlements can occur, although this positive causal effect must sometimes overcome the reluctance of highly centralized power to allow movement of settlement boundaries outward and, hence, farther from direct monitoring and control. In addition to the administrative infrastructure, centralized power will initially encourage production, which then has effects on markets, transportation technologies, and the distributive infrastructure which expand the scope and scale of the material infrastructure and encourage immigration, thereby increasing the size, density, and expansion of settlements. But as power becomes highly centralized, it begins to dampen production and technological innovation in ways that stagnate or decrease the immigration and infrastructural development that fuel the size, density, and expansion of settlements.

Second, as power becomes centralized, inequality increases and, as a consequence, internal threats escalate. These cause further centralization of power around administration and coercion, which work back to influence infrastructural processes revolving around distribution, material facilities, and administration. At first, such centralization may encourage the expansion of production, technological innovation, and markets, but in the end, it will dampen these forces and, as a result, the development of the infrastructural base that encourages immigration and settlement growth. Growth will become less dynamic; and urban space will increasingly come to reflect the needs of the administrative infrastructure for control rather than the synergy that comes with less-regulated relations among technology, production, distribution, and infrastructural development. Moreover, as power becomes centralized in response to internal threats, even more repression and tight regulation occur, with the ironical consequence of increasing these threats and undermining the legitimacy of centers of power. In so doing, the potential for disintegration is increased. Such disintegration, or even the perception of its eventuality, will increase rates of migration, some of which is initially away from volatile urban centers but much of which will eventually be back to urban areas as disintegrative processes intensify and disrupt life in both rural and urban areas.

Third, as power is centralized and used to conquer and control territories, rates of migration increase. For as people's lives are disrupted, they move—sometimes away from settlements but more often into urban areas in search of opportunities. As they do so, the size and density of settlements increase. Expansion of territories also escalates second-order logistical loads as increases in the absolute size of a territory and the ethnic/cultural diversity of the population within this territory ratchet up the disintegrative potential. Such increased potential will cause further centralization of power which, as noted above, has a positively curvilinear effect on those infrastructural forces that increase size, density, and expansion of settlements.

Aside from these paths of geopolitical influences on settlement patterns are those portrayed on the left side of the model. All of these exert positive effects on the infrastructural development that increase the size, density, expansion, and agglomeration of settlements. These positive effects can be dampened, or even reversed, as a result of the overcentralization of power (and hence, the effects of centralized power on production and distribution as well as on inequalities and internal threats).

Forces affecting settlement patterns are, reciprocally, very much influenced by the settlement patterns that they set into motion—which is why I believe that we must view spatial processes a separate macrodynamic force in their own right. First, large settlements make it possible to support a larger population, primarily as a result of development in the material infrastructure as this is caused by those forces affecting the distributive and administrative infrastructures. Second, large, dense, expanded, and agglomerated settlements have reverse causal effects—some direct, others indirect—on increasing the scope and scale of the distributive, material, and administrative infrastructures which, in turn, cause increases in the volume, velocity, and scope of markets, in the level of consolidated power, and in the level of production. Third, large, dense, expanded, and agglomerated settlements can increase the level of internal threat and, hence, cause power to be more centralized or disintegration to be more imminent by virtue of creating an "urban mass" that can be mobilized in movements of counterpower. Thus, both directly and indirectly, settlement patterns have an independent set of causal effects on other macrodynamic forces—population, production, distribution, power, and, as we will see shortly, differentiation.

Elementary Principles of Spatial Dynamics

The model presented in figure 6.3 is obviously complex, but by translating key causal paths into propositions, it is possible to simplify the analytical model and generate testable hypotheses. Propositions I, II, III below state the basic forces determining, respectively, the size of the territory controlled by a population, the size, density, and expansion of settlements within that territory, and the overall level of agglomeration of settlements within that territory.

I. The size of the territory controlled by a population is a positive function of its success in war with other populations, which, in turn, is a positive function of:

A. the relative level of coercive power, which is a positive and additive function of:
1. the size of a population
2. the level of production
3. the level of material surplus
4. the level of consolidated and centralized power
B. the extent of a marchland advantage

II. The size, density, and expansion of settlements within the territory controlled by a population are a positive and additive function of:
A. the scale and scope of the material infrastructure, which, in turn, is a positive and additive function of:
1. the scale and scope of the distributive infrastructure, which is a positive and additive function of:
a. the level of communication/transportation technologies
b. the level of production
c. the volume/velocity/scope of markets
d. the extent/rate of geographical expansion of settlements
e. the scale/scope of the material infrastructure
2. the scale and scope of the administrative infrastructure, which is a positive and additive function of:
a. the level of consolidated power
b. the level of centralized power
c. the level of second-order logistical loads
d. the scale/scope of the administrative infrastructure
3. the level of production, which is:
a. a positive function of first-order logistical loads stemming from population size
b. a positive curvilinear function of the level of centralized power
c. a positive function of the volume/velocity/scope of markets
4. the volume/velocity/scope of markets, which are:
a. a positive function of the level of production
b. a positive function of the level of first-order logistical loads

c. a positive curvilinear function of the level of consolidated power

d. a positive function of the scale/scope of the distributive infrastructure

B. the net rate of immigration, which, in turn, is a positive and additive function of:

1. the level of transportation/communication technologies

2. the scale/scope of the distributive infrastructure, which is a function of the conditions listed under II-A-1

3. the degree of success in war

4. the diversity of a population

5. the size/density of existing settlements

6. the level of disintegrative pressures

III. The overall level of agglomeration of settlements is a positive and additive function of:

A. the extent/rate of geographical expansion of settlements, which is:

1. a lagged positive function of size/density of settlements, which is a function of the conditions listed under II above

2. a positive function of the scale/scope of the material infrastructure, which is a function of the conditions listed under II-A

3. a positive function of the scale/scope of the distributive infrastructure, which is a function of the conditions listed under II-A-1

4. a positive function of the volume/velocity/scope of markets for real estate

5. a positive function of the scale/scope of the administrative infrastructure, which is a function of the conditions listed under II-A-2

B. the scale/scope of the distributive infrastructure (see II-A-1)

C. the scale/scope of the administrative infrastructure (see II-A-2)

D. the volume/velocity/scope of markets (see II-A-4)

7

DIFFERENTIATION DYNAMICS

SOCIAL DIFFERENTIATION, or the creation of distinctive types of organizational units, social categories, and systems of cultural symbols within a population, has long been considered the basic macro structural process. Early social theorists—from Adam Smith to Auguste Comte through Herbert Spencer to Émile Durkheim—all viewed differentiation as the central problematic of their theories; and more recently, neofunctionalists have continued the emphasis on differentiation as the central social process (e.g., Alexander 1985; Alexander and Colomy 1990; Colomy 1990; Luhmann 1982).

Yet social differentiation is, in a sense, an analytical construct for describing what we have already discussed in previous chapters. For underlying the dynamics of population, production, distribution, power, and space is differentiation; and we might question whether or not a separate discussion of differentiation is necessary. Yet my sense is that such a discussion is useful because differentiation is not just an outcome of the processes described in previous chapters, it also has independent effects on population, production, distribution, power, and spatial organization. Spencer ([1874–96] 1898) was the first to recognize this fact when he argued that differentiation is more than a response to logistical loads; it also provides a "structural skeleton" that enables populations to grow, production to expand, distribution to escalate, power to consolidate and centralize, and territories to be transformed.

Even if one disagrees with this conclusion, a chapter on differentiation can serve as a convenient way of providing a provisional summary of previous chapters. Because differentiation is involved in population, production, distribution, power, and spatial dynamics, an analysis of this underlying process provides an opportunity to pull together conceptually the argument developed thus far.

In developing this chapter, I will emphasize two modes of theorizing that, ultimately, owe their inspiration to biological analogies: (1) functionalism, where organismic analogies have produced an emphasis on those processes determining the differentiation and integration of systemic wholes, and (2) human ecology, where Darwinian analogies[1] have led to a concern with competition, selection, and differentiation of social units. Although these two traditions remain distinctive, their initial founders—Herbert Spencer ([1874–96] 1898) and Émile Durkheim ([1893] 1933)—used both in developing their macro structural theories of social differentiation. In this one-hundred-year period since the seminal analysis of Spencer and Durkheim, however, functional-organicist theorizing has maintained its emphasis, indeed an overemphasis, on the integrative forces holding differentiating populations together,[2] whereas both urban[3] and organizational[4] ecology have sustained a focus on competition and selection as a driving force behind differentiation of the structures organizing a population but have failed to link these dynamics to the larger social whole within which they occur.

Early Convergence of Ecological and Organismic Analyses

Spencer's Demographic-Geopolitical Model

Even though the Darwinian-sounding phrase—"survival of the fittest"—can be found in early philosophical works by Spencer ([1851] 1888), much of his sociology ([1874–96] 1898), save for the discussion of political institutions, is devoid of such metaphors.

In figure 7.1, I have recast earlier models on Spencer to emphasize the demographic and geopolitical forces affecting differentiation. For Spencer, population growth is the primary cause of differentiation,

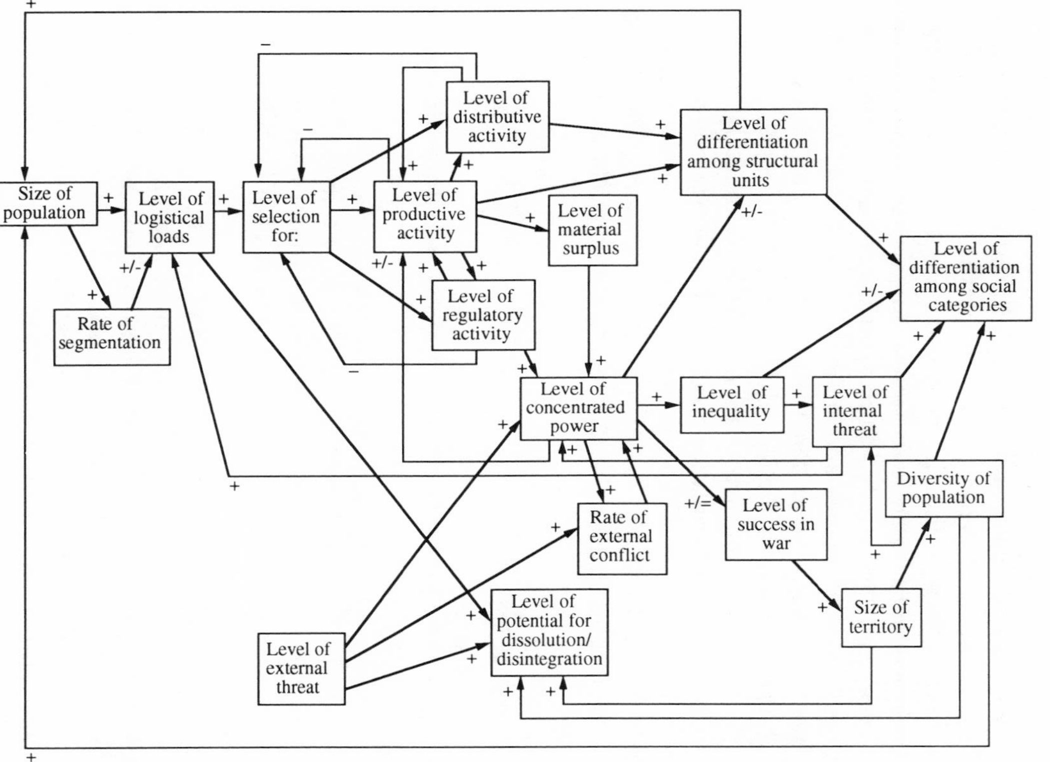

Figure 7.1. Spencer's Demographic-geopolitical Model of Differentiation

although he recognized that reverse causal processes in productive, regulatory, and distributive axes of differentiation can become forces of differentiation in their own right. Moreover, these axes have effects on inequality which, along with structural differentiation, generate distinctive social categories. Added to these forces is a geopolitical one that intensifies the concentrations of power and inequality, which, in turn, initially increase the number of social categories until inequality becomes so great as to polarize categories into the "haves" and "have-nots." Successful geopolitical activity also generates larger territories and more-diverse populations, which also increase categoric differentiation directly and, indirectly, through the reverse causal effects delineated in the model.

Spencer's picture of differentiation is perhaps limited because of the overemphasis on population size as the prime macro-level force, and perhaps, because of the concern with geopolitics. But it is a suggestive model and gets us started in conceptualizing some of the forces involved in structural and categoric differentiation.

Durkheim's Ecological Model

Durkheim's ([1893] 1933) macro structural theory of differentiation is more clearly ecological because structural differentiation is viewed as a function of the competition and selection processes delineated in the middle portion of the model in figure 7.2. For Durkheim, differentiation occurs on (1) a structural level (numbers of different positions and, by implication, the organizational units in which these positions are lodged) and (2) a cultural level (systems of symbols involved in regulation of positions and organizational units). Differentiation at both levels is the result of density and competition as these are affected by population size and technologies that reduce the "spaces" between individuals. Structurally, differentiation increases the selection pressures for cultural differentiation, as the reverse causal arrow from differentiation back to intensity of selection underscores, but in Durkheim's mind, differentiation also decreases competition and exerts a counterpressure on intensity of selection. Out of these selection pressures come diverse systems of cultural symbols, some highly abstract evaluative symbols (e.g., values) and others more specific and regulatory (e.g., laws, norms). If such a system of differentiated symbols can be developed, then they feed back and enable (indeed, encourage) further structural differentiation because a joint system of abstract values and other evaluative symbols provides diverse actors with a common perspective, while more specific regulatory symbols facilitate the coordination among diverse units and, as the model indicates, the formation of networks among structural units.

Thus, Durkheim's model introduces the Darwinian metaphor into sociology—that is, density, competition, and selection—and connects these processes to "social speciation" or specialization in the division of labor and to cultural forces that integrate differentiated units in order to avoid "anomie" (or lack of regulation). Durkheim assumed that anomie would eventually be corrected by selection pressures causing the differentiation of cultural symbols—an assumption that was perhaps more wishful thinking than reasoned analysis. Yet Durkheim provided some critical leads to a theory of social differentiation.

Functional and Ecological Views on Selection Processes

In looking back at Spencer's and Durkheim's models, many of the key elements of a macro-level ecological theory can be found. What makes their theories ecological is the notion of "selection," but this concept denotes somewhat different processes in Spencer's and Durkheim's theories. As I emphasized in chapter 2, one type of selection is Darwinian or Durkheimian in deference to Durkheim's adoption of Darwinian metaphors: differentiation is related to the number of resource niches, the scarcity of resources in niches, and the level of competition and selection *among* social units or idea systems in these niches. Another type of selection is organicist or Spencerian in recognition of Spencer's insight: differentiation is related to disintegrative pressures in social wholes and selection *for* certain types of social units and symbol systems as a means for avoiding disintegration. In the Durkheimian view, selection operates under conditions of high

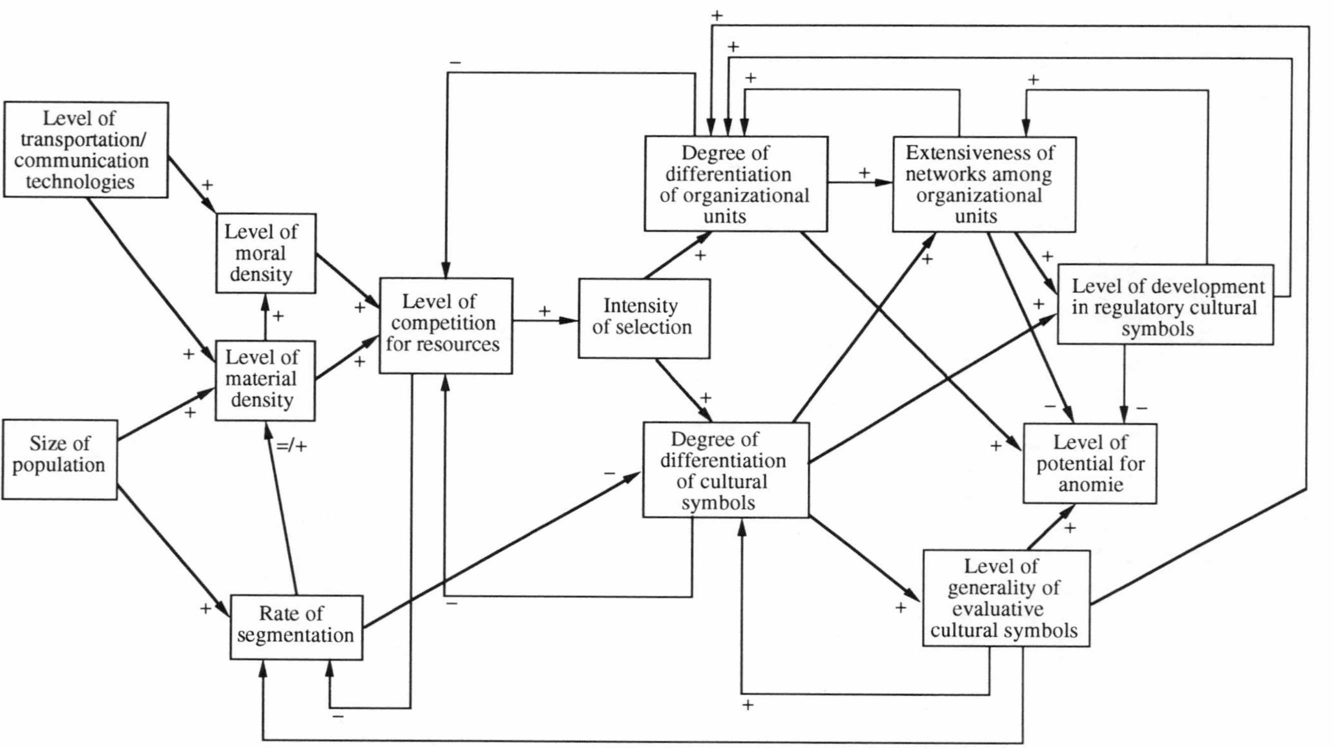

Figure 7.2. Durkheim's Ecological Model of Differentiation

density in a niche; in the more organismic approach—or what I termed "Spencerian selection"—there is a condition of low density, or even an absence of structures and symbols that can promote integration in the face of increased logistical loads and pressures for disintegration. For Spencer, there are non-Darwinian selection processes operating in the differentiation of a societal population; these involve selection *for* structural units that can resolve escalating logistical loads for regulation, production, and distribution as well as the loads created by inequality, large expanses of territory, and diversity of population. For Durkheim, there are also non-Darwinian processes revolving around selection for layers of regulatory and evaluative symbol systems that can enable subgroups, organizational units, and individuals to coordinate their activities and form integrative social networks.

This distinction between Durkheimian and Spencerian selection makes a great deal of difference in how macro-level analysis is conducted. If one emphasizes Durkheimian selection, then the key processes are resource levels, resource niches, niche density, competition for resources, and selection *among* units; as a consequence, other macro-level elements, such as population, production, distribution, power, war, territory, symbol systems, inequalities, and technology, are viewed in terms of how they affect resource levels, niches, density, and competition. In contrast, if one stresses organicist or Spencerian selection, the key processes are those disintegrative forces that generate selection *for* structures and symbols that can overcome these disintegrative forces; and as a result, ecological processes revolving around niches, density, and competition are given less emphasis. Although present-day human ecology was initiated by the macro-level functional theories of Spencer and Durkheim, it has developed independently of the functional concerns of early theorists; and over time, ecological theory has emphasized its affinities to bioecology (e.g., Hawley 1944, 1950, [1971] 1981; Hannan and Freeman 1977). This divergence of ecological and macro-functional theories has been useful in that the metaphorical and vague Darwinian arguments articulated by Durkheim and implied by Spencer have been made considerably more precise, although they are still analogies rather than homologies.

The Divergence of Ecological and Functional Theorizing

A General Model of Organizational Ecology

As I did in chapter 6 on spatial processes, where I extracted and abstracted the urban ecology model, it is useful to extract the general model of organizational ecology.[5] For the critical node of structural differentiation is an organizational unit, and hence, theories on the ecology of organizations should be relevant to macro-level theory. The model in figure 7.3 draws upon a number of sources[6] and seeks to integrate them for my macro-theory purposes.

The basic line of argument is Durkheimian: when resources are abundant, the number of organizations grows to the point where density increases and sets into motion competition and selection among organizations. But unlike Durkheim, ecological theory stresses that organizations can fail and are selected out as a consequence of being outcompeted by other organizations in the same niche, or invading organizations from other niches. In some versions of organizational ecology (e.g., Hannan and Freeman 1977), organizations develop inertial tendencies which have the effect of giving selection a stable structural "phenotype" to work on. But others would argue in a more Durkheimian vein that organizations do not have to be selected out; rather they can change and seek new resource niches (McPherson 1981, 1983a, 1983b, 1990).

Thus, organizational ecology presents two images: one in which organizations are selected out because of structural inertia under conditions of high density and competition, and the other, in which organizations can change, move, and adapt to different resource niches. The latter position presents an argument of how organizations become differentiated: under conditions of competition, organizations restructure themselves and seek new kinds of resources elsewhere. Moreover, they may even create new resource niches in which they can survive. In contrast, those viewing organizations as revealing inertial properties see organizations of a given type as increasing in number and, then because of structural rigidities, as being selected out under competition. But from a macro position, we need to ask: Why do they die out from density alone? Should not

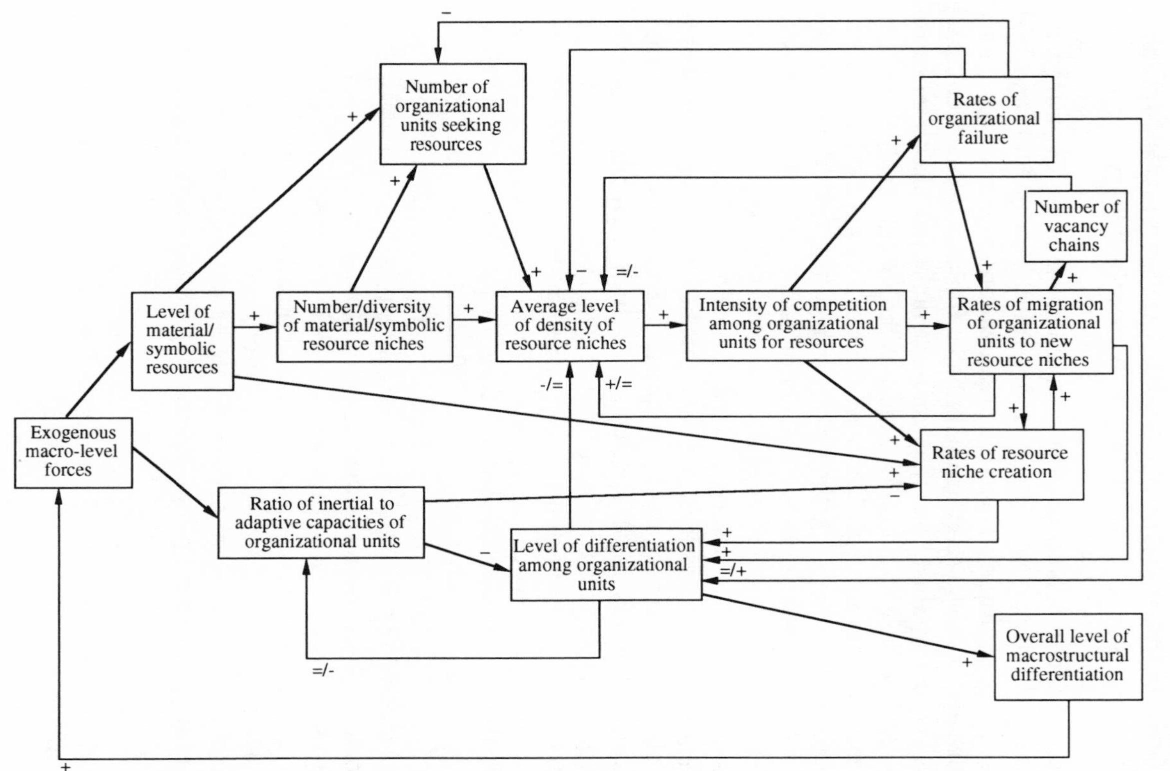

Figure 7.3. The Abstracted Organizational Ecology Model of Differentiation

an equilibrium point be reached at which organizations and resources come into balance? Why is it that goal-oriented and reflexive units cannot change, finding new sources of resources? The answer to these questions resides in understanding the macro forces that affect resource levels and diversity, coupled with what organizations inside and outside a niche do under conditions of high and low density. For example, as outside organizations invade a niche, they change and, hence, become differentiated from what they were; similarly, less-fit organizations that can flee a niche will alter their structure and increase the differentiation at the macro level; and either successful invading organizations or those fleeing a niche can create new resources on which they can survive. The net effect of these processes, as organizations move about, is to increase niche density initially, but once high rates of movement occur, a sufficient number of vacancy chains is created so as to reduce niche density.

These kinds of questions and issues are not central to much organizational ecology because the emphasis is on meso- rather than macro-level processes, although theorists such as Hannan and Freeman do mention government activity and market processes, whereas others, such as McPherson, stress population size and network density as important macro-level forces. These kinds of forces become highly relevant when one shifts attention to questions such as: Where do resources and niches come from? What determines their level and diversity? What are the rates of migration of organizations into and out of niches? These questions are largely exogenous to much organizational ecology, with notable exceptions, such as McPherson, but they are central to using this perspective in macro-level theorizing. For clearly, macro-level forces such as population size and diversity (by categoric distinctions), levels of production, scale and scope of distribution, and degree of consolidated and centralized power all operate as resources, both material and symbolic, in which organizational units can survive. The model in figure 7.3 emphasizes processes endogenous to the organizational dynamics that are set into motion as a consequence of increased density and competition in resource niches. But our goal will ultimately be to connect these meso-level theoretical ideas to theories emphasizing exogenous, macro social processes.

Reconciling Meso-Level Ecological and Macro-Level Functional Analyses

Hawley's Macro-Level Theory

Amos Hawley's (1986) more recent ecological theory begins a reconciliation of macro-level functional and ecological theories. In figure 7.4, Amos Hawley's[7] core ideas on the dynamics of macro structural differentiation are remodeled with an emphasis on social differentiation. Hawley (1986) introduces what an essential distinction in macro structural analysis between "corporate units" and "categoric units"—one pursued by McPherson (1990) in his discussion of "Blau-space," whereby the differentiated characteristics of population members become distinctive resource niches for organizations. This emphasis was, of course, anticipated by Spencer's brief portrayal of social categories. Differentiation of corporate units involves creating distinctive types of organizational structures, whereas differentiation of categoric units revolves around the emergence of social distinctions that use symbols to classify individuals.

Hawley sees mobility costs, which are determined by transportation/communication technologies, as a critical consideration because they set limits on how much differentiation and specialization of organizational units can occur. For without the capacity to move information, material, and actors, differentiation is not possible (a line of argument anticipated by Spencer's emphasis on distribution). Increased production directly causes differentiation of corporate units because of the efficiencies associated with specialization; and this process is accelerated by the existence of markets, competition, and selection among specialized units which seek to find a stable and predictable resource niche. Population size, as it encourages expanded production and as it provides a larger social mass which can be divided in more ways, also works to increase in the number and size of diverse units. Indeed, like Spencer, Hawley recognizes that larger populations must be divided into a greater number of organizational units if they are to be sustained. Yet if population growth raises mobility costs in the absence of new

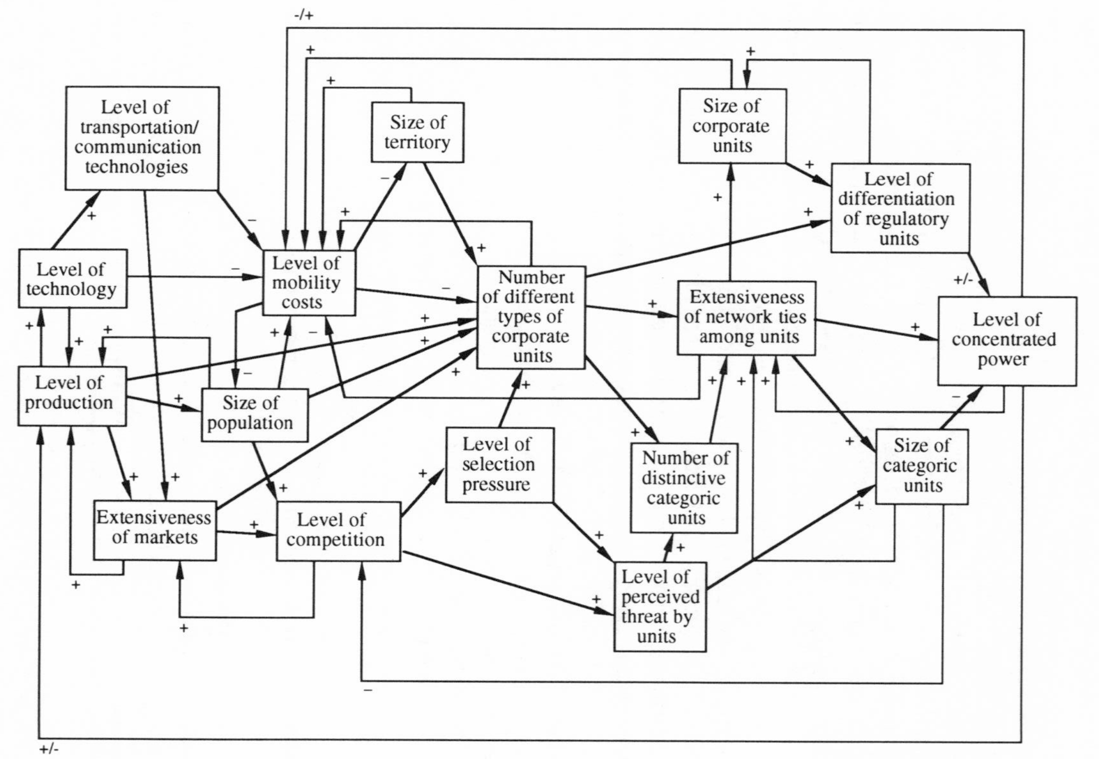

Figure 7.4. Hawley's Macro-level Model

technologies and increased production, then it works against differentiation of corporate units. Differentiation of corporate units and growing size of a population, along with the competition and threat generated by production and markets, increase the number and size of categoric units. In turn, the existence of many diverse types of corporate and categoric units creates selection pressures for their control through the differentiation of regulatory units (i.e., concentrations of power) and through the consolidation of units in increasingly extensive networks. The concentration of power at moderate levels encourages further differentiation through its effects on mobility costs and capital formation in the productive sector. In turn, increased production causes further differentiation of corporate and categoric units, expanded numbers of network ties, increased efforts at regulation, and increased concentration of power (which eventually will work against increased production).

These theoretical leads in Hawley's (1986) scheme consolidate, to some degree, the ideas in Spencer's and Durkheim's functional approaches, while adding refinements. Yet surprisingly, this work moves more toward functional concerns than ecological ones, analyzing differentiated units in terms of "functions" and "key functions" (Hawley 1986) for each other and for the maintenance of a system in its environment. As such, the scheme downplays somewhat Hawley's earlier emphasis on resource niches, competition, and selection (e.g., Hawley 1944, 1950, 1973). Indeed, it moves toward technology as the driving force behind differentiation, along the lines of Lenski's (1966) and other evolutionary theories, but it does so by stressing transportation and communication technologies and their effects on mobility costs. Thus, the unit of the analysis has shifted in Hawley's work from the meso level (i.e., communities and organizations) to the macro level (i.e., societal populations). In some ways, these shifts weaken the ecological component, but at the same time, they point to avenues for reconciling ecological and organismic (i.e., functional) theories.

A Synthetic Model of Macro-Level Differentiation

In figure 7.5, a model reconciling the organicist selection arguments of functionalism with the Darwinian selection analogies of human ecology are pulled together with much of what has been said in earlier chapters and from sources whose ideas seem pertinent. Again, as I noted in the last chapter, the model is complex not as a result of descriptive intent but more out of my desire to cast a broad conceptual net in visualizing the general forces and their causal connections that operate to produce differentiation. The model can be decomposed into smaller chunks, as I will demonstrate shortly, but as a starting point for understanding differentiation it is useful to visualize generic processes and their many points of intersection in their most complete form.

In the figure, there are three sets of selection processes highlighted: those on the left pertaining to first-order Spencerian selection, those in the middle revolving around Durkheimian selection, and those on the right pertaining to second-order Spencerian selection. These three blocks of selection processes will organize my discussion of the model in figure 7.5.

First-Order Spencerian Selection Processes

At the left of the model are a series of forces that set into motion Spencerian selection processes. These are termed *first-order* because they involve selection *for* those basic structures needed to stave off potential dissolution of the population. Perhaps this emphasis is only a rewording of old functional arguments with all of their problems (Turner and Maryanski 1979), but they are nonetheless fundamental to understanding the social organization of societal populations.

Most of these selection processes were delineated by Spencer in *The Principles of Sociology* ([1874–96] 1898), but I have added refinements suggested by others. With Spencer, I argue that first-order selection is initially caused by population growth. For when a population grows, several forces are set into motion as a consequence of the increasing logistical loads for sustaining the population: (1) As Durkheim stressed, segmentation of like structural units is one response to growth; and as Hawley would argue, I think, such is particularly likely to be the case under conditions of low technology and high mobility costs (as denoted by the reverse causal arrow from mobility costs to segmentation). (2) As McPherson (1990) has argued,

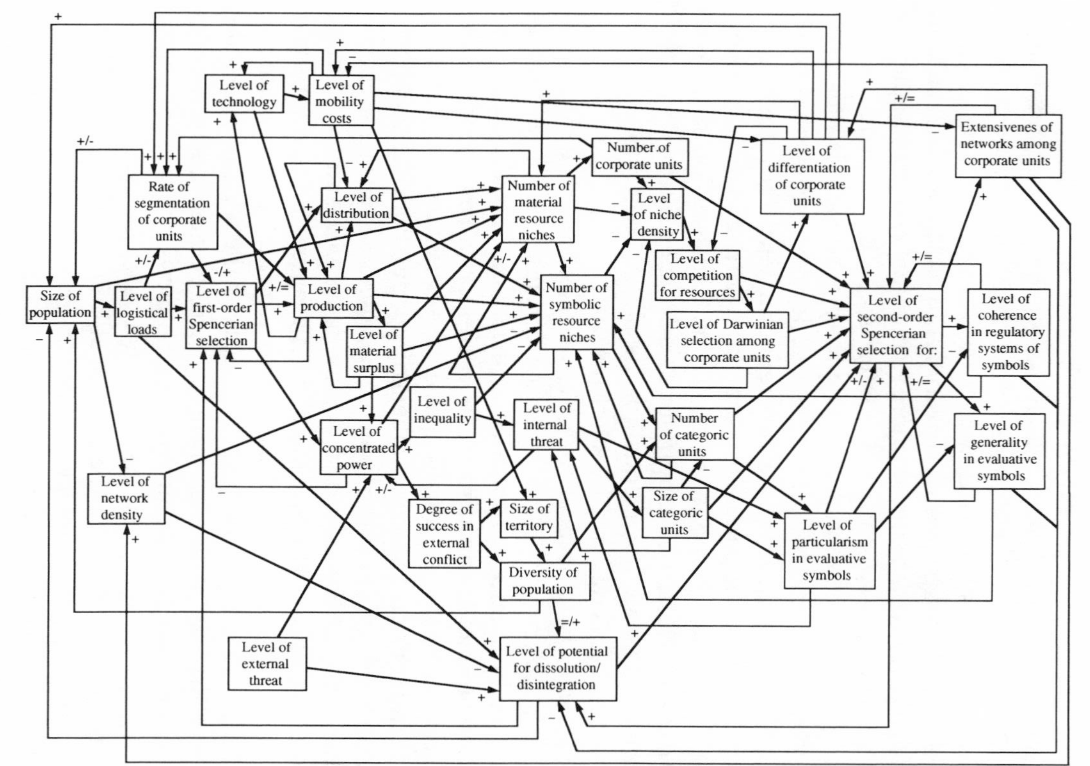

Figure 7.5. A General Model of Differentiation Dynamics

population growth decreases network density among members of a population, thereby decreasing capacities for social control and, as Spencer (1862, [1874–96] 1898) believed, setting into motion forces for movement and localization of sub-units and subpopulations in new resource environments, which cause either their differentiation from one another or their dissolution as a coherent system (such is possible, Durkheim averred, when there are few ecological constraints maintaining material density). And as Mann (1986) or Maryanski and Turner (1992) have more recently argued, the "caging" of populations creates pressures for internal conflict and competition leading to differentiation, but the capacity of subpopulations to escape ecological and political cages also leads to differentiation by virtue of their localization in diverse environments. (3) Segmentation under conditions of population growth eventually becomes inadequate as the sole response to mounting logistical loads, as both Spencer and Durkheim emphasized; as a consequence, logistical loads escalate, and the potential for dissolution increases.

Together, as is denoted by all of the direct, indirect, and reverse causal arrows into the level of first-order selection, these forces generate selection *for* expanded activity along the three principle axes suggested by Spencer: production, distribution, and regulation. The level of production must increase to support a larger population; distribution processes must expand to sustain the movement of goods and services among members of a population; and regulation through the concentration of power must increase in order to coordinate and control the increased volume of activities among a larger population. For as I stressed in previous chapters, if these outcomes are not forthcoming, then mounting logistical loads cause the dissolution of the population. Thus, it is at the level of first-order selection that population is a primary force because it sets into motion many other forces that will become increasingly more important. Again, let me stress that I am not asserting a demographic determinism, but I do feel that the significance of population forces has not been given sufficient attention in general theory.

Once expanded production, distribution, and regulation (via the consolidation and centralization of power) are set into motion, they become forces in their own right and are less driven by population forces than by their own internal dynamics and by the other forces that they, in turn, have set into motion. The first cluster of these forces is delineated in the left-middle portion of the model in figure 7.5: (1) As Hawley recognized and Durkheim implied, increased production is both the result and cause of technological development, which, in turn, lowers mobility costs as new transportation and communication technologies are developed. (2) Lowered mobility costs facilitate distribution, which then stimulates more production and technological development. (3) Expanded production increases material surplus, which can be used as capital to expand production further, but as Spencer and more recent scholars such as Lenski (1966) have emphasized, it is also usurped by centers of power and used to increase the concentration of power as well as by elites to form a basis for new kinds of categoric distinctions. (4) As power becomes concentrated, inequality escalates as Spencer and modern-day evolutionary-conflict theorists document (e.g., Lenski 1966; Turner 1984b); and as a consequence, internal threats increase and, ironically, lead to greater concentration of power in order to control such threats (up to the point where internal threats are so high as to create viable centers of counterpower). (5) As Spencer and more contemporary theorists such as Lenski (1966) and Collins (1986) have recognized, the concentration of power is often related to geopolitical processes, because external threat increases concentration of power in order to mobilize, control, and channel resources to deal with the threat; and in the long run, these processes aggravate the cycles increasing internal threat. Moreover, if external conflict ensues, and society is successful in war, then the growing amount of territory, size of population, and diversity of this population all escalate logistical loads and first-order selection for increased power, production, and distribution.

Durkheimian Selection Processes

The forces that set into motion by first-order Spencerian selection processes generate the resource niches within which selection *among* corporate and categoric units takes place. That is, Spencerian selection *for* increased production, distribution, power, and their effects on technology, mobility costs, inequality, and geopolitics determine the level and varieties of resources available for Durkheimian selection among structural units organizing members of a population.

This outcome is the result of the obvious fact that to increase the values for these processes causes segmentation and differentiation of structures and categories, which, in turn, become resource niches for the Durkheimian selection processes emphasized by ecological theories. These Darwinian-like dynamics are delineated in the middle-right portion of figure 7.5. Increases in production, distribution, and concentrations of power directly cause the number of resource niches to expand, although highly concentrated power and property will begin to decrease the political resources available to organizational units and, via the feedback loop to material resources, the material resources as well. These same forces will also expand the number of symbolic resource niches via the paths delineated in the model.

A clear tautology is evident in these generalizations, however. Increased levels of productive, distributive, and political activity involve the very differentiation that is to be explained by these levels. For most meso-level organizational ecologists, this tautology is not apparent because resource levels as generated by production, distribution, and power are, for the most part, exogenous variables or unexplained constants. But when we seek to account for why resources exist at a given level and variety, creating resource niches that allow for differentiation of corporate and categoric units, we come face to face with the argument that differentiation of resource-generating processes produces differentiation.

This tautology is part of reality: segmentation and then differentiation create niches for subsequent differentiation and segmentation; and in turn, this further differentiation increases the levels of those forces—production, distribution, and power—that generate more resources in existing niches while increasing the number of niches and expanding the resource base for segmentation of differentiated units. Thus, differentiation is a self-escalating process, up to the levels that technology allows and up to the point where overly concentrated power and inequality in property work directly and indirectly (via causal processes delineated in the model) to decrease both the diversity and levels of resources. In the model in figure 7.5, this self-escalating nature of differentiation is delineated by numerous reverse causal processes that directly and indirectly lead back to affect the number of material and symbolic resource niches, which then set into motion the Darwinian selection processes outlined by organizational ecology.

Turning first to the key causal processes revolving around material resources that are used by corporate units, several processes can be highlighted: (1) As the arrows flowing into this variable emphasize, the number of material-resource niches is affected by human bodies (as members, clients, participants), money, products, goods, and other units of value that can be used to sustain a structure, physical objects and space, and coercive capacities, although these are almost always mingled together in some combination to produce a variety of resource niches. (2) As McPherson has emphasized (McPherson and Ranger-Moore 1991), symbolic resources defining attributes of individuals in Blau-space (i.e., resource niches defined by the attributes of actors and their distribution) also interact with material resources in creating niches for corporate units; and as Spencer and Hawley recognized, inequalities and internal threats create categoric units that can become highly volatile and dynamic resources for corporate group formation (e.g., as a social movement or organized revolt). (3) As all ecological theories stress, the number of corporate units of a given type will expand to fill an underexploited material-resource niche, a process which is a form of segmentation that fuels, via the long feedback loop to segmentation and logistical loads, selection for those processes that increase the level and diversity of resources. Thus, segmentation is essential to differentiation by virtue of its effects on increasing niche density. (4) As the level of density increases, the degree of competition for resources escalates, leading to Darwinian selection among corporate units whose inertial tendencies give selection something to work on and whose adaptive abilities may allow them to readjust an existing resource niche, to find and invade a different niche, or to create a new niche. And, (5) as competition and selection escalate, the level of differentiation among corporate units increases, feeding back to increase the number of resource niches and, in turn, encouraging further segmentation of corporate units in a niche.

Market dynamics (as part of distributive processes) accelerate the amount and diversity of material resources. As I implied in chapter 4, markets are the critical mediating structure between technology, production, and material surplus, on the one hand, and the creation of resource niches on the other. Because markets connect individual preferences and demand to production, they can differentiate to meet the virtually unlimited preferences of individuals; and in so doing they can create "markets for" virtually any good, service, or symbol, which, in turn, can become resource niches for corporate units. As a result, markets can dramatically expand the number of niches, often without increases in aggregate production or material surplus (White 1981). For if markets are relatively open, their existence encourages sellers to seek "new markets," including the marketing of the terms of exchange (e.g., money, futures, options, etc.) of lower-level markets; and in so doing new resource niches are generated (Collins 1990). As a consequence, once markets reach a threshold of differentiation and evidence fiscal mechanisms (money, credit, insurance, etc.) that lower the mobility costs of transactions, they tend to expand and differentiate rapidly, up to the point where (*a*) the resource niches created by such differentiation become overexploited, (*b*) the level of speculation in such niches has produced goods and services beyond the level of demand, thereby setting into motion a market collapse, or (*c*) the level of demand for goods and services levels off, or declines. Capitalist markets take these dynamic qualities to the extreme, but they are inherent tendencies in all market systems that reach a certain level of differentiation. As the limits of (*a*), (*b*), and (*c*) above are reached, markets often collapse, thereby dramatically reducing the level of resources and the diversity of resource niches available to structural units.

Turning now to the key causal processes revolving around symbolic resources, several can be highlighted: (1) the number of symbolic resource niches is potentially greater than material resource niches, up to the limits of the number of individuals in a population and their imaginative capacities to create niches in Blau-space. (2) Production, distribution, and power all directly increase symbolic niches by virtue of the division of labor's partitioning effects on incumbents' experiences, a process that is expanded indirectly by the creation of material resource niches. (3) Lowered network density directly increases the number of symbolic niches because of its effects on lowering interconnections among actors (both their "moral" and "material" density in Durkheim's terms), thereby reducing the effects of social control to enforce conformity to common symbols. (4) Concentrated power as it effects inequality and ethnic diversity (through geopolitics) is also a driving force behind the formation of symbolic niches. (5) The number of categoric units is related to not only the number of symbolic niches but also the level of material resources (at a minimum, human bodies who are categorized but typically other material resources such as money, property, power, place in division of labor, etc.) necessary to sustain a category in Blau-space. (6) The size of categoric units expands as a result of internal threats, which are fueled by inequality and ethnic diversity. (7) Categoric units, in turn, increase the symbolic resources available for corporate units. And, (8) the ecological dynamics of competition and selection among categoric units are, I argue, related to the density of corporate units in symbolic resource niches; and so, selection among categoric units occurs primarily when they have become wholly, or partially, imbedded in corporate units within densely populated symbolic resource niches.

In sum, then, Durkheimian selection among corporate units and, indirectly, categoric units connected to corporate units, is circumscribed by first-order Spencerian selection processes, but via a number of reverse causal effects delineated in the model in figure 7.5, such Durkheimian selection alters the course of these Spencerian selection processes. One major reverse causal effect emerging from Durkheimian selection is, as Spencer was most apt to emphasize, the capacity of a more differentiated set of corporate units to support a larger population and to organize this population in ways that respond to first-order functional selection pressures, thereby avoiding disintegration of the population. Another related causal effect occurs via what I term in the model in figure 7.5 *second-order Spencerian selection processes.*

Second-Order Spencerian Selection

The processes delineated on the right of figure 7.5 are second-order Spencerian selection in this sense: the consequences of Darwinian selection within a population create a new level of escalated logistical loads revolving around how the differentiated corporate units are, themselves, to be integrated so as to avoid dissolution. This problem was initially given its most forceful expression by Adam Smith ([1776] 1805), but it was soon incorporated into the French lineage culminating in Durkheim ([1893] 1933).

We could argue, of course, that the process of differentiation creates resource niches for integrative structural units and symbol systems (e.g., laws, courts, enforcement, etc.), but as with first-order Spencerian selection, we can go only so far because we are addressing selection *for* sociocultural processes under conditions of low (or lack of) density among such integrative structural units. In broad contours, I think that Durkheim ([1893] 1933:329–89, 1902) understood these processes best as a series of simultaneous pressures for (1) the formation of larger networks of interdependence among corporate units, which, as Hawley (1986) also recognized, is related to mobility costs, (2) the generalization of evaluational systems, or what Parsons (1966) later termed "value generalization," and (3) the development of coherent regulatory symbols for (*a*) attaching generalized values to concrete situations and (*b*) articulating rights, duties, and obligations among corporate units. All of these processes are self-reinforcing because each generates selection for the other: generalized systems select for regulatory symbols to fill in the relevant details (in order, in Durkheim's eye, to avoid "anomie"); regulatory symbols select for value premises to give them meaning and moral force; and similarly networks of organizational interdependency select for both specific and generalized symbols to regulate transactions among units. Of course, there is no guarantee that these selection processes will be successful; indeed, in the long run, they have failed for most human societies. Thus, we must avoid the functionalist trap of assuming that cultural integration will occur, just because it is needed. In fact, often when it is needed, it is too late, because conflict groups have developed their own particularistic symbols.

Increases in the level of inequality, the number of categoric units, and the size of such units (as fueled by internal threats) thus create particularistic evaluational symbols which work against generalized symbols and coherent sets of regulatory symbols. For inequality always creates structural divisions and symbols reflecting these divisions; and these more particularistic symbols (among classes, ethnic groups, religious groups, regional subpopulations, and the like) work as a counterforce to consensus over either regulatory or generalized symbols. At the same time, however, these particularistic symbols escalate second-order Spencerian selection pressures for new regulatory and generalized symbols—often without result, however. For particularistic symbols can overwhelm a population, subverting second-order Spencerian selection and setting into motion dissolution and disintegration of a population.

Elementary Principles of Differentiation

Figure 7.5 also emphasizes that the ecological analysis of macro structure requires a reunification of the selection arguments in both analogies—organismic and Darwinian—from biology. As long as human ecology operates at the meso level of analysis, such reunification is not necessary, but as Hawley's (1986) more macro turn suggests, functional ideas need to be reintroduced when larger-scale societal processes are examined from an ecological perspective. In broad strokes, the specifics of ecological analysis—the level of material and nonmaterial resources, the number of niches, the density of niches, the competition for resources, and the intensity of Durkheimian selection—are variables whose values are determined not only by the indigenous dynamics of Durkheimian selection itself, but also by the outcomes of Spencerian selection for certain structures and symbols; and conversely, the outcomes of first- and second-order Spencerian selection are circumscribed by the Durkheimian selection processes that they set into motion. At the first-order level, Spencerian selection initiates forces—population growth, expanded production, market distribution, and consolidation of power—that, via their effects on each other and on stratification and geopolitics, influence the levels and kinds of resources in niches where Durkheimian selection among corporate units operates. At the second-order level, differentiation of corporate units generates an additional set of forces—subnetworks, regulatory symbols, and generalized evaluative symbols—that, through complex reverse causal chains, feed back to exert further influence on both first-order Spencerian selection and Durkheimian selection among corporate units.

In way of extracting key social processes from the model in figure 7.5, let me conclude by offering some preliminary propositions:

I. The overall level of differentiation of a population is a joint and positive function of:

A. the number of material resource niches in that population, which, in turn, is:

1. a positive function of population size
2. a positive multiplicative function among the levels of production, distribution, and material surplus, which are:
    a. a positive function of the level of first-order Spencerian selection pressures
    b. a positive function of the level of technology
    c. a negative function of mobility costs
3. a positively curvilinear function of the relationship between the consolidation and centralization of power, which is:
    a. a positive function of first-order Spencerian selection
    b. a positive function of the level of material surplus
    c. a positively curvilinear function of internal threat
    d. a positive function of external threat
4. a positive function of the previous level of differentiation among corporate units, which is:
    a. a positive function of the level of Durkheimian selection as such selection encourages an increase in adaptive to inertial tendencies in organizational units
    b. a negative function of mobility costs
5. a positive function of the number of symbolic resource niches, which is a function of the conditions listed under I-B below

B. The number of symbolic resource niches in a population, which, in turn, is a positive and additive function of:

1. the multiplicative relation between the levels of production and distribution, which is a positive function of the conditions listed under A-2 above

2. the number of material resource niches, which is a positive function of the conditions listed under A above
3. the level of previous differentiation of categoric units, which is a positive function of:
   a. the size of categoric units
   b. the number of symbolic resource niches
   c. the diversity of the population
4. the level of coherence in regulatory symbols and the level of generality of evaluative symbols, which are a positive function of second-order Spencerian selection
5. the dampening effects of network density among members of a population

8

DISINTEGRATION DYNAMICS

THE ORGANIZATION OF A population in space and over time involves an often-precarious standoff between forces of disintegration and integration. Yet the terms "integration" and "disintegration" have problematic connotations in sociology, often being fused with ideological, as opposed to scientific, questions. Such need not be the case, however, as long as we recognize that these terms are merely labels for sets of processes enabling a population either to sustain itself as a coherent whole in its environment or to become less coherent and viable. Whether integration or disintegration is good or bad becomes irrelevant; the real theoretical issue is conceptualizing how these processes operate.

As the chapter title emphasizes, I see disintegration as the more likely end state of all populations. Integration is only temporary; for throughout history, populations have overcome disintegrative forces, building up integrative structures and systems of symbols, only disintegrate in the long run. Disintegration does not mean that a population in a territory ceases to exist, although such has been the case many thousands of times over in the long run of human history; rather, disintegration denotes the fact that a population has lost its coherence vis-à-vis its environment, at least for a time.

I also emphasize disintegration because it is necessary to avoid the pitfall of functional theories that have stressed integrative forces. This emphasis of functional theory on integration is not entirely misplaced, however, for sociology is about how populations organize themselves; and the notion of social organization must involve assessment of integrative forces that sustain a pattern of organization. But it is also critical to view these integrative forces as only temporary, rarely lasting more than a few hundred years in sustaining a population as a distinctive whole in its environment. And when this long-term evolutionary view is taken, we should be quick to recognize that populations continue to disintegrate today and that even those seemingly stable and unassailable political democracies in the contemporary world will, in the long run, disintegrate. We must recognize, of course, that integration is a crucial dimension of the social universe, but it is the forces of disintegration that ultimately override those processes holding a population together.

Integrating Forces

Integrating forces hold members, and the units organizing them, together, so that the population as a whole does not evidence high rates of conflict, deviance, and other dissociative processes that would lessen its coherence and its viability vis-à-vis other populations as well as physical parameters of the environment. To some degree, the forces of disintegration are the converse for those promoting integration, but disintegration involves much more, and so I think it best to view integration and disintegration separately.

There are two levels on which integrative forces operate: (1) the structural, which concerns the patterns of relations within and between units organizing a population; and (2) the cultural, which denotes the systems of symbols that are used to sustain patterns of relations within and among organizing units. Let me begin with the structural forces.

Structural Forces of Integration

The forces causing integration emerge and persist by virtue of Spencerian selection pressures—that is, by selection *for* solutions to disintegrative potential. Such Spencerian selection generates pressures to find solutions to disintegrative forces, or face the consequences. The search for solutions often fails or, given the circumstances, cannot be implemented (Tainter 1988). And so, in the end, all societies disintegrate, but for the moment, our concern is with those structural patterns that stave off this inevitability. What are they?

Segmentation

The most basic form of structural integration is segmentation, or the creation of new and similar structural units to alleviate mounting logistical loads. As Durkheim ([1893] 1933) and Spencer ([1874–96] 1898) were perhaps the first to conceptualize theoretically, the reproduction of similar units enables a population to grow without differentiation, thereby eliminating the problems reconnecting diversely structured units. Instead, the existing organization of positions and associated cultural symbols are simply reproduced in ways that attach individuals to replicas of the same basic type of structural pattern. Moreover, in attaching individuals to similar structural forms and systems of associated cultural symbols, congeries of what Sailer (1978) terms "regular equivalence" are sustained, with individuals in equivalent positions behaving and thinking in convergent ways. Because of equivalence, segmentation can integrate large numbers of individuals because as "copies" (or roughly so) of new organizational units are spun out, the patterns of social relations and systems of cultural symbols remain the same and, as a result, individuals in one unit have little difficulty in understanding and adjusting to the responses of those in another unit. From another angle, DiMaggio's and Powell's (1983) discussion of "institutional isomorphism" makes a similar argument, but the more general point is that segmentation operates as an integrative force, at least up to a point. Additionally, segmentation can work to reduce the formation of distinctive sub-units with their own particularistic cultural systems that challenge the general values and normative symbols of a population.

Even as differentiation increases in a population, segmentation still operates as an integrative force as similar structural units and associated symbols are generated along major axes and nodes of differentiation. For example, there may be considerable differentiation of business corporations in capitalist societies, but those in similar sectors of the economy will be structured in roughly equivalent ways and will generate similar symbols. Furthermore, even if there is clear differentiation among such corporate bodies (say, between a mining company and a retail clothing store), there is a certain broad equivalence in their structural forms as private corporations seeking a profit; and as a consequence, their members will display convergent behaviors and commitments to common symbols. Thus, segmentation still operates as an important integrative force even as differentiation accelerates.

Segmentation is eventually supplemented by differentiation as direct connections among structural units decline (i.e., the overall network density decreases) with the growing size of a population, with increasing segmentation, with expanded territories, and with development of those forces, such as production, distribution, and power, that population growth and geopolitical activity often set into motion. As Spencer was the first to recognize, such differentiation creates a more elaborate "structural skeleton" that can support a larger social mass; and as all functional theorists (who ultimately took the problem from Adam Smith) have stressed, differentiation generates selection pressures for new patterns of structural interdependence.

Structural Interdependence

Some degree of interdependence, or the mutual exchange of people, materials, and services, is evident in segmentally structured systems (exchanges of brides across kin-based villages being a common example in horticultural societies). But with greater degrees of segmentation and, eventually, differentiation among units, exchanges increasingly involve unlike resources. That is, organizational units come to depend on the particular resources of another unit; and as chains of such interdependence are built up through exchanges, some degree of integration is achieved. Such exchanges are sustained by what Emerson (1962, 1972) termed "balancing operations," where actors come to accept the rate of exchange in resources, even if they are somewhat disadvantaged and dependent.

Market processes transform these dynamics in complex ways. First, they increase the scope of exchanges and enable vaster chains of interdependence to evolve, pulling more units together into relations of direct and indirect exchange. Second, markets increase the velocity of exchanges, moving resources more rapidly among units. Third, markets escalate both the volume and diversity of resources connecting units. Fourth, markets eventually create a highly liquid and, hence, usable generalized medium—money—that can rapidly define value in exchanges and, as a consequence, increase the aggregate level of value among units engaging in exchanges. Fifth, markets expose alternative options to actors involved in exchanges and thereby facilitate the "balancing" of exchanges (Emerson 1962, 1972) since actors have been able to search alternatives and can now accept as the "best option" a particular exchange relation (by adjusting expectations for value or utility), even if they are disadvantaged in the exchange. And sixth, such a medium neutralizes more affective and traditional media in ways that can reduce, though rarely eliminate, internal conflicts stemming from particularistic values, such as ethnicity, regionalism, or productive specialty.

Structural Inclusion and Embeddedness

Differentiation produces units of varying sizes, and in so doing, it generates larger units in which smaller units are embedded. Such inclusion of units within units promotes integration by defining and delimiting the terms of exchange relations among units. The nature of inclusion can vary from some degree of overlap through more cooptive control of one unit by another to hierarchic control of smaller units within a larger one.

Inclusion mitigates some of the disintegrative tendencies of exchange interdependence because it limits "negotiation" over, and "shopping around" for, alternatives; and in so doing, the potential tension and conflict inherent in more-open exchanges are reduced. Instead, inclusion usually involves a clear definition of obligations, an accepted medium or media of exchange, and an institutionalized flow of resources among units. If inclusion is hierarchic and rigid, however, it can create rigidities in units that escalate the level of conflict.

Structural Domination

The consolidation and use of power are, of course, central mechanisms of integration. Power has the capacity to control and coordinate the actions of individuals and structural units, and it becomes essential to the viability of a large population, as has been consistently emphasized in previous chapters. Power is most effectively used when leaders have consolidated all of its bases—the administrative, coercive, symbolic, and material—and rely on a combination of material incentives issued through an administrative structure that is co-optive, legitimate, and noncoercive. Such balance has, however, rarely been sustained and, as we will see, power is also one of the chief disintegrative forces within a population.

Indeed, consolidation and centralization of power tend to be self-perpetuating: centralization of power breeds more centralization and, thereby, begins to throw out of balance the bases of consolidation, in several ways. First, overregulation through inflexible administrative structures escalates resentments. Second, centralized power increases inequalities that cause resentments to rise and symbolic bases of legitimation to erode. Third, overregulation by centralized authority decreases mobility, positional opportunities, and other forces breaking down particularistic categoric units and structural rigidities. Thus, domination is always a precarious balance between under- and overregulation; too little causes chaos, whereas too much generates resentments and undermines other integrative processes, such as cross-cutting affiliations and positional mobility.

In a sense, the creation of political democracy is a response to this dilemma of domination. By enabling more broad-based participation by the public in decision making, or by creating at least the illusion of such participation, oscillations between over- and underregulation need not erode legitimacy. Indeed, diffuse legitimacy for the structure of power can be sustained at the same time that particulars of decision making are rejected. In this way, potential mobilization by categoric units with particularistic symbols that challenge the entire structure of power is more likely to be channelled toward the specifics of grievances rather than toward the legitimacy of power itself, thereby reducing the disintegrative force of mobilized categoric units. Indeed, such mobilization often operates a mechanism for releasing tensions that could accumulate into more radical opposition (Coser 1956, 1967). But, just whether democracy is a permanent solution to the problems inherent in domination is an open issue, despite some scholars' liberal faith in the end of history. For democracy has its own contradictions, the most important being: (1) the public and interests constantly put pressure on political elites to deal with new sets of problems and, in so doing, encourage the expansion of regulatory bureaucracies at the very time that the public's beliefs advocate less regulation; and (2) the openness of democracies to "interest politics" can encourage the proliferation of interest groups and public activism to the point where decision making becomes deadlocked, thereby creating pressures for more regulation (and perhaps less democracy) or forcing an increasing amount of decision making to be made in the bureaucratic structures of the state. Thus, it is not clear that the democracies which have evolved over the last few hundred years are a permanent "solution" to problems of integration. They have been extremely effective in promoting integration through power, but we should not be too ready to view them as a final solution to the problems inherent in the consolidation and use of power. But no matter what the historical outcome, the fact remains that power and domination are a critical integrative force in human populations.

Cross-cutting Affiliations

To the degree that incumbency in diverse organizational units is uncorrelated, facilitating membership in varying configurations of organizational units, integration is promoted. Peter Blau (1977) emphasized this in his notion of "intersecting parameters," but the argument is essentially one of cross-cutting affiliations. In particular, to the degree that members of what Hawley termed "categoric units," especially categoric units formed around inequalities of resource distribution, can also belong to organizational units uncorrelated to categoric membership, integration is promoted. For if memberships in hierarchies and categories become consolidated (Blau 1977), systems of particularistic cultural symbols emerge and inclusion/embeddedness works to produce potential conflict groups as members have little knowledge or experience with those outside their groupings and as stereotypical portrayals of these "others" emerge.

Market processes do much to promote unconsolidated parameters by enabling individuals to use a neutral and nonparticularistic medium to pursue diverse needs and interests. For as McPherson (1988, 1990) emphasized, even if money does not actually change hands, market-driven systems generate competitive resource niches whose organizations seek new members, thereby creating diverse opportunities that can lower the degree of overlap or consolidation of membership in categories and corporate units.

Positional Mobility

Movement of people, and even organizational units, generally promotes integration, unless such movement threatens individuals and corporate units (as is the case with immigrants to new areas). Yet even when movement initially creates disintegrative pressures, contact and interaction can over time generate new bonds and networks, while decreasing particularism and parochialism.

Movement across space, or up and down vertical hierarchies, thus promotes integration by decreasing partitions among social categories and corporate units, while increasing rates of interaction and network formation across categories and corporate units. Although tensions and conflict often ensue, at least initially, such mobility decreases the probability that the population will polarize around consolidated parameters, to use Blau's (1977) terms, involving high correlations among memberships in categoric and organizational units.

Social differentiation, per se, increases mobility by virtue of expanding the array of positional opportunities and by generating vacancy chains as people and organizations move about (White 1970). Production is particularly important in generating mobility because it creates new resource niches for the organizations providing positional opportunities. Market volume, velocity, and expansion all create new positional opportunities, but equally important, these forces constantly shift demand for goods and services which, in turn, shift positional opportunities as new resource niches are created and as new categoric distinctions emerge. Moreover, by virtue of shifting the dominant medium of exchange to money and away from particularistic considerations, markets diminish the hold of particularistic cultures on individuals and, at the same time, create a new, more universalistic set of symbols that encourage movement to new places and positions (because the "terms" of incumbency will more likely be more universalistic).

Power also can affect mobility, but in a more curvilinear manner. Initial consolidation and centralization of power create new positional opportunities themselves, but they also provide entrepreneurial services facilitating growth and differentiation in production and markets. But if power becomes too centralized, it overregulates and stagnates production and distribution, thereby decreasing positional opportunities. Moreover, highly centralized power tends to increase inequality, which decreases rates of mobility and fosters the formation of disjointed, if not polarized, categoric units that pull and consolidate corporate units in ways lessening the intersection of parameters and cross-cutting relations.

Structural Segregation

If potentially conflicting subpopulations are separated by their patterns of organizational affiliations and/or by their distribution in space, such a situation promotes, for a time, integration. But, in the long run, it will increase disintegrative pressures through its effects on the consolidation of parameters among categoric and corporate units, thereby decreasing mobility and discouraging the formation of cross-cutting affiliations. And, if such segregation is associated, or consolidated, with inequalities in the distribution of resources, then it will become a potentially volatile force of disintegration. As power is centralized to cope with such disintegrative potential, its use can initially be effective in sustaining segregation, but the use of power inhibits differentiation and mobility in all sectors of a population, while increasing the very inequalities that make segregation a volatile force. Thus, at best, segregation is only a temporary integrative process; in the end, it is a disintegrative.

In figure 8.1, I have modeled the causal processes emphasized in the preceding discussion. Rather than summarizing each causal arrow, let me simply highlight particular clusters of causal effects.

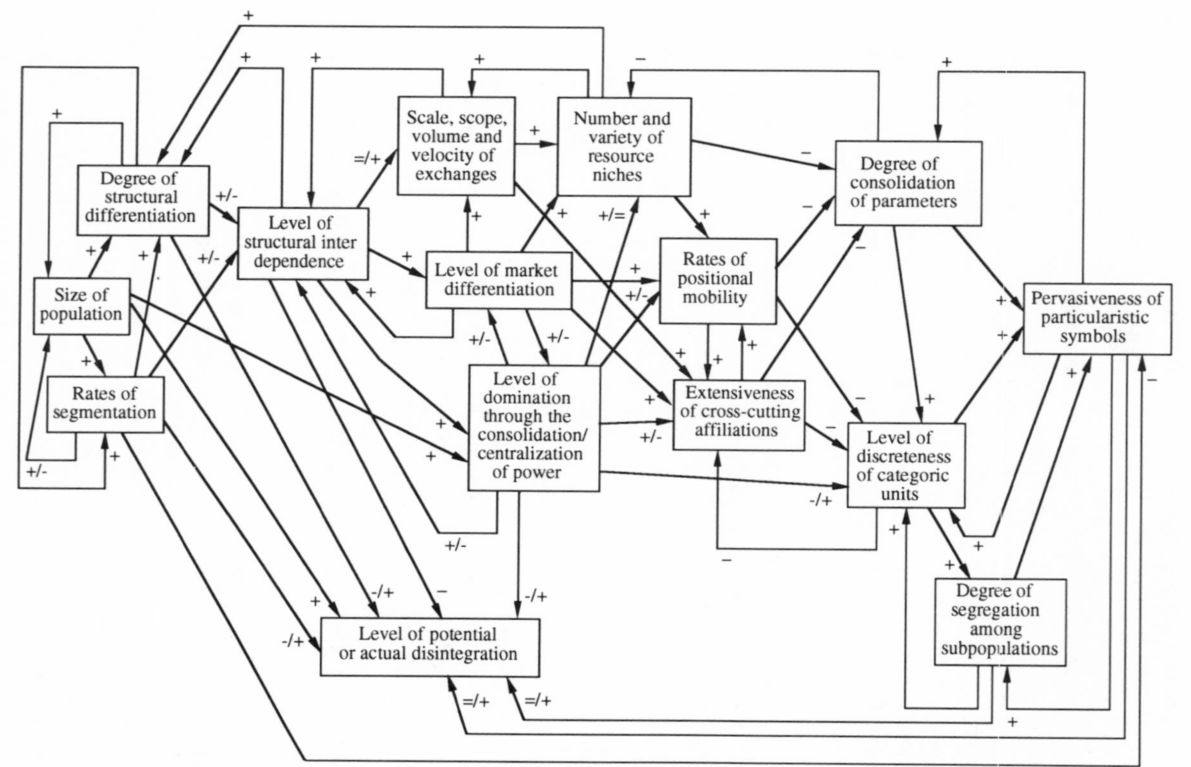

Figure 8.1. The Dynamics of Structural Integration

Structural differentiation, as fueled by population growth, is integrative when it leads to structural interdependence revolving around: (*a*) market differentiation, increased rates of exchange, and increased numbers of resource niches; (*b*) domination based upon a balanced consolidation of power and moderate degrees of centralization; (*c*) high rates of positional mobility; and (*d*) high levels of cross-cutting affiliations, or intersecting parameters. In contrast, to the degree that structural differentiation does not lead to (*a*) through (*d*) above and, instead, to (*a*) highly centralized power, (*b*) consolidated parameters, (*c*) the emergence of particularistic sets of symbols, (*d*) discrete categoric units, and (*e*) prolonged segregation of subpopulations, such differentiation will increase disintegrative potential.

Just how these integrative and disintegrative processes operate is, to some degree, contingent on the circumstance and history of a population. If, for example, a society is created by conquest or revolution, the high degree of centralized power necessary to sustain order will increase the disintegrative potential of this population, whereas if a society is formed by relatively free immigration and unfettered by past conquests and internal conflicts, there is more integrative potential. But the basic relations among the forces of structural integration are unchanged; only the values for each are altered by contingent circumstances and history.

Cultural Forces of Integration

Symbols are a powerful and, obviously, necessary dimension of human organization; indeed, via intense Spencerian selection pressures, they can become an integrative force which can, to a degree, compensate for a lack of structural integration. And yet selection can also work to generate particularism and conflict group formation, leading to disintegrative pressures. For the moment, however, let me focus on the ways that symbols can promote integration.

Generalized Evaluational Symbols

As Durkheim ([1893] 1933) emphasized, structural differentiation causes evaluational symbols to become highly generalized, a process that Talcott Parsons (1966, 1971) later termed "value generalization." Although it has become somewhat chic, in both functional (e.g., Luhmann 1982) and conflict (e.g., Collins 1975) circles, to deemphasize the significance of generalized cultural standards for assessing right, wrong, good, bad, and other dimensions of morality, the emergence of these generalized standards is essential to macro structural integration. True, as Collins (1981, 1988) and other theorists advocate, these symbols are reinforced and sustained by interactions in local encounters, but more is involved: once actors possess them, they have a life of their own; they are not just epiphenomenon because, independently of specific structures and encounters, they provide a set of common and generalized reference points and guidelines for action by both individuals and collective actors. And in so doing, they regulate actions and the formation of structures. And without the abstraction and generalization of such evaluative symbols, the integration of differentiating populations becomes problematic.

Even if there is consensus among such symbols, however, their attachment to diverse social contexts always poses a problem. As Luhmann (1982) has argued, ideology becomes the intermediate symbol system between highly generalized values, on the one hand, and positions in organizational units in various functional domains (e.g., economy, polity, law, kinship, science, education, medicine) on the other. I prefer the term "beliefs" to ideology because of the latter's connotations; and in my view (Turner and Starnes 1976), there are two levels of beliefs attaching values to institutional domains: (1) evaluative beliefs which translate generalized values to institutional contexts and specify what should occur in these contexts and (2) empirical beliefs that represent cognitions about what actually exists in an institutional domain, although such "factual" interpretations are highly distorted by evaluative beliefs. For example, a generalized value such as "achievement" becomes an evaluative belief ("work hard") in the economic sphere, which biases empirical beliefs ("opportunities and rewards go to those who work hard").

Rarely is there a situation in which considerable slippage among generalized values and various levels of beliefs does not occur. But if some degree of correspondence cannot be sustained, then integration at a macro structural level becomes problematic. Moreover, these layers of evaluative symbols can overcome structural pressures for disintegration, such as is the case when those in low-ranking categories and positions believe that "with hard work" they can be mobile (often an unrealistic belief, but one that mitigates against the formation of conflict-oriented particularistic beliefs that attack the culture and structural arrangements of a population). Without this capacity of systems of evaluational symbols to, in Marx's terms, systematically generate "false consciousness" among its most disenfranchised members, a population will experience dramatic increases in disintegrative pressures.

Yet generalized values can become a source of disintegration through conflict when actual structural arrangements deviate too far from the premises of evaluational symbols. Under these conditions, the evaluational symbols become a criterion for assessing situations and potentially a basis for discrediting existing structures and for opposition group formation. Thus, the level of correspondence between values and structural arrangements becomes a crucial condition of integration; and beliefs help establish this correspondence by: (*a*) translating evaluational symbols into more concrete, mundane, and less potentially volatile symbols, and (*b*) generating a certain amount of false consciousness that hides the inevitable lack of correspondence. Still, when there is too much slippage between generalized evaluational systems and structural arrangements, such abstract symbols can be used to mobilize opposition groups.

Moreover, as Parsons tended to emphasize, "universalism" accompanies the generalization of evaluational symbols. Such universalism offers the sense that evaluational symbols are applied consistently, fairly, and equally to all; and if this illusion can be sustained, then the universalism inherent in value generalization promotes integration. Yet as with all abstract value premises, universalism is a double-edged sword; it can become an unrealistic yardstick by which actual structural arrangements are assessed and, then, questioned. And because the everyday operation of social structures inevitably involves particularism and because truly unlimited application of universalistic standards would probably bring most operations to a halt (because of the bureaucratic and legal entanglements involved in assuring "fairness"), there is always the potential for conflict group mobilization. Again, beliefs can operate effectively to place boundaries on universalism and to sanction "appropriate" particularisms in ways that deflect this potential.

Another ironic twist to generalized and universalistic claims is that they often become the basis for particularistic symbols and categoric as well as corporate group formation. By asserting a general ethical principle, derived from existing evaluational symbols, and asserting it to be applied to all, conflict groups enshrine their particularistic symbols in the rhetoric of universalism. For example, Christianity was universalistic in its willingness to accept all, but the history of Christianity has involved a constant splintering of Christian organizational units that develop more particularistic beliefs that are couched as even more universalistic than other Christian organizational units. Similarly, highly differentiated populations tend to generate corporate groups making universalistic claims—for example, animal rights, environmental harmony, equal rights, rights to life, rights to control one's body, etc.—which fuel the intensity of particularistic beliefs that mobilize such groups to pursue conflict. Simmel ([1902] 1950) was perhaps the first to recognize, at least by extension of his ideas, that the "blasé" character of actors in highly differentiated systems comes from a cynical and somewhat alienated stance from the ever more shrill and unrealistic claims of universalism behind the proliferation of particularistic causes. At some point, Simmel implied, a psychological numbness and unwillingness to be pulled into the ideological fray set in; and I should add that this is a powerful integrative force. Because if members of a population can live with constant ideological conflict, then this willingness promotes integration. Thus, just as universalism may contribute to conflict group mobilization in terms of particularistic symbols (that are asserted to be universal), it contributes to the ability by members of a population to tolerate relatively high rates of group conflict.

Regulatory Symbols

As Durkheim ([1893] 1933) recognized and as Niklas Luhmann (1985) has recently reconceptualized, a system of regulatory rules is essential to integration among differentiated units, especially in populations where evaluative symbols are generalized and where the volume and velocity of transactions are high. Following his mentor, Talcott Parsons (1966, 1971), Luhmann stressed that an autonomous legal system, capable of self-reflection over procedures and precedents, and designed to produce positive law that can be altered with changing conditions, is essential to integration (Fuchs and Turner 1991).

Such laws become increasingly necessary with differentiation of basic institutional spheres—that is, economy, power, family, education, religion, science, medicine—from one another and with market mechanisms increasingly mediating exchanges and movements among spheres. For without rules regulating transactions, disintegration is likely. Moreover, an autonomous legal system can generate rules and procedures that mitigate against overregulation by centers of power, and the disintegrative pressures thereby generated. Without a separate system for creating regulatory symbols, interests in the economy, polity, or religion cannot be controlled, and as a result, conflict and increased potential for disintegration will ensue.

Still, an autonomous legal system can also give rise to conflicts by providing resource niches for litigation and protest movements. In so doing, it can mitigate the conflicts (Dahrendorf 1959), or at least release some of the tension over them before they lead to disintegrative violence (Coser 1956, 1967). Indeed, when generalized and universalistic evaluational symbols encourage contact group formation in terms of particularistic symbols, an autonomous legal system can defray much of the intensity of these groups. But, on the other side of the issue, the existence of such an autonomous legal system can encourage increased rates of confrontation and, more importantly, can cause such an ever escalating system of rules that the complexity of the system, itself, begins to create logistical loads and a new source for disintegrative potential. Yet without such a system of law, the degree of structural differentiation is limited because of the incapacity to mediate relations among the differentiated units.

Legimitating Symbols

Sets of symbols must legitimate the rights of key actors in the productive, distributive, and power arenas. If such actors' rights to engage in activity are questioned, then disintegrative tendencies increase.

Legitimacy operates, I believe, at two levels: (1) a diffuse level in which highly generalized evaluative symbols bestow rights on key actors to exist and engage in activity and (2) a specific level in which diffuse symbols are translated into sets of regulatory rights. If only diffuse symbols exist (as is usually the case, for instance, in monarchial systems), then abuse of bestowed rights eventually occurs, thereby creating disintegrative propensities. If only specific legitimating symbols exist (such as a Constitution and body of tort law), then the underlying commitment to the rights of actors is weakened. Durkheim's ([1893] 1933) discussion on "the moral basis of contract" emphasized this fact, because regulatory symbols without a value-laden system of generalized symbols lack sufficient moral force to be sustained as an effective source of legitimacy.

Generalized Media

George Simmel's ([1907] 1978) insightful analysis of money as a medium that transforms the structure of social relations was elaborated upon by Talcott Parsons (1963a, 1963b, 1970) and later Niklas Luhmann (1982). What these elaborations emphasize is that transactions within key institutional spheres are conducted in terms of systems of symbols, which provide (*a*) the symbols necessary for discourse, (*b*) the symbols for engaging in nonverbal interaction, (*c*) the symbols for defining the units and terms of exchanges, (*d*) the symbols for developing beliefs that connect evaluational and regulatory symbols, and (*e*) the symbols for legitimating the activities of key actors.

With increased differentiation, especially of distributive-exchange processes in complex systems of markets, money becomes the most significant generalized medium for macro-level integration. For all of the effects listed above, but especially (*a*), (*c*), (*d*), and (*e*), money increasingly becomes the mediator of transactions and exchanges within and between actors in institutional spheres, while being the object of discourse, the basis of beliefs, and the cornerstone of legitimacy (particularly in the economic, distributive, and political arenas). The reason for this is that money has two unique qualities.

First, money is a relatively neutral medium that limits the development of particularized systems of symbols. As a medium, money mitigates against, and in fact often destroys, the domination of symbols that promote the formation of categoric units whose members construct their own distinctive media which partition subpopulations from each other. Instead, money can be used across organizational and categoric boundaries, and in so doing, it promotes mobility and affiliations among actors in differentiated units. Still, as is evident in the world today, money is never completely successful in mitigating categoric unit formation and the formation of subpopulations with distinctive symbols, but it works against such formation. Yet if money is unstable (i.e., hyperinflation), then it can encourage the development of particularistic symbols. This occurrence only goes to demonstrate, however, the integrative effects on money.

Second, money is a medium that increases value, as Simmel ([1907] 1978) emphasized. When money is spent in a transaction, it is used to purchase value. And because money is generalized and usable in many diverse contexts, it provides actors with many options for increasing subjective value. Thus, money has the capacity to increase aggregate subjective value in a population, and in so doing, it binds actors to structures rather than alienating them, as Marx and Engels ([1846] 1947, [1848] 1978) incorrectly emphasized. Yet when the value of money is unstable, then it has the opposite effect that is more in line with Marx and Engels's projections. Additionally, in the spirit of Marx and Engels, money comes to denote the value of property and, hence, to give a more precise measure of inequality, which, in turn, can escalate deprivations by those without property and push them to organize for conflict.

These two qualities of money, its liquidity and convertibility as well as its capacity to increase aggregate value, make it a very important legitimating force. When actors can increase value and their options for doing so, they bestow diffuse legitimacy on centers of power. Of course, if the inflation of money decreases options (because other actors will not accept money in exchanges) and lowers its value (because of its declining purchasing power), then it becomes a potent force of delegitimation of centers of power, while increasing (indeed, stimulating) the reliance on more particularistic symbols that will tend to increase disintegrative pressures.

In sum, then, systems of cultural symbols, operating as evaluational premises, as regulatory rules, as legitimating processes, and as generalized media of exchange, are essential to integration. Conflict theories have, I think, underemphasized these processes, and as a consequence, they have failed to see these cultural processes as the reason why conflict does not always occur. For cultural symbols have the capacity, I argue, for mitigating potentially problematic relations of exchange and interdependence, inclusion, and domination; and they facilitate positional mobility and the formation of cross-cutting affiliations while counteracting the particularizing effects of structural segregation. Moreover, as rational-choice theories would emphasize, the cost to actors of conflict and of "free-riding" are high; and under these conditions, it is more utility-maximizing to construct symbols that regulate conflict and free-riding (Hechter 1987; Coleman 1986, 1990; Heckathorn 1990). But still, having said this, culture can only go so far; there are always powerful disintegrative forces in the organization of human populations.

It is, of course, difficult to disentangle cultural from structural forces, but in figure 8.2, I have sought to do so. Along with Durkheim, and later Parsons and Luhmann, I argue that structural differentiation, especially with respect to what Spencer conceptualized as "operative" (i.e., production, reproduction), distributive, and regulatory functions, sets into motion Spencerian selection for the forces enumerated in the model.

As Adam Smith ([1776] 1805) first argued and as Durkheim ([1893] 1933) reemphasized, differentiation causes values to become highly generalized so that they can encompass the increased diversity of activities and experiences of actors. Once such generalization occurs, there are intense Spencerian selection pressures for attaching value premises to activities and the structural units within which, and between which, such activities occur. This attachment occurs along several dimensions: (*a*) the development of evaluative and empirical beliefs corresponding to value premises; (*b*) the emergence of legitimating beliefs, both diffuse and specific, for activities and structures in functional domains, especially productive, distributive, and regulatory; and (*c*) the evolution of regulatory symbols as these emerge from the differentiation of an autonomous rule-making structure (i.e., a legal system).

These forces interact with another set of forces unleased by struc-

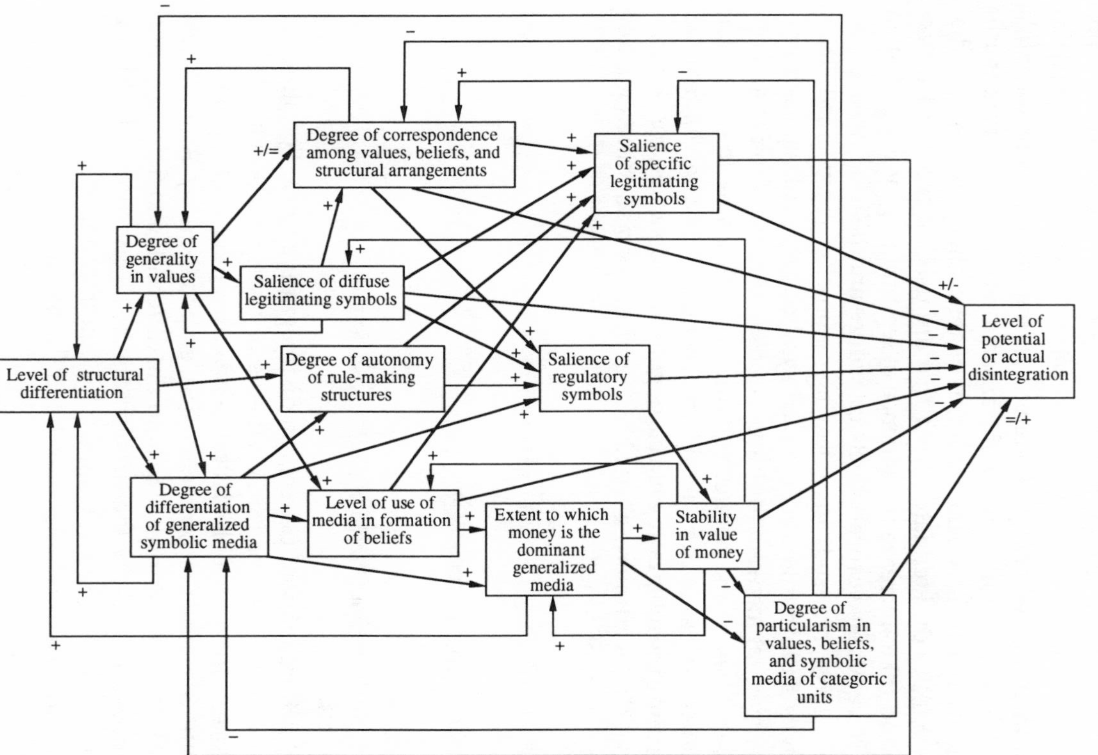

Figure 8.2. The Dynamics of Cultural Integration

tural differentiation: generalized symbolic media. As Parsons and Luhmann argued, distinctive media evolve to facilitate transactions in differentiated functional domains, and once these exist they enable what Luhmann (1982) termed "self-thematization" of activities in a domain, or what I have viewed as the development of beliefs about the appropriate nature of transactions within and among domains. Such beliefs, couched in terms of generalized symbolic media, operate (*a*) to differentiate media that are used to conduct transactions and (*b*) to increase the salience of legitimating symbols that, in turn, have reverse causal effects on values and beliefs.

As I have stressed, in following Simmel's ([1907] 1978) lead, money is an important generalized symbolic media because it provides a neutral medium for transactions within distributive and operative processes and between all functional domains. In so doing, it mitigates against particularistic values, beliefs, and media that can undermine the integrating effects of generalized values, corresponding beliefs, legitimating symbols, and functional differentiation among symbolic media.

Disintegrating Forces

The Disintegrative Effects of Differentiation

Social differentiation, per se, is a disintegrative force. For although differentiation may be initially necessary, via Spencerian selection processes, to reduce logistical loads resulting from population growth, it poses the problems first emphasized by Adam Smith ([1776] 1805): What force is to hold the differentiated units together? This question became the problematic of all functional theorizing, beginning with Auguste Comte (1830–42) through Herbert Spencer ([1874–96] 1898) to Émile Durkheim ([1893] 1933) and well on into the twentieth century with A. R. Radcliffe-Brown (1952) and Talcott Parsons (1951). The integrative processes discussed above constituted the corpus of their answer to this question. Yet these forces of integration must confront counterforces of disintegration that are always inherent in structural differentiation. Let me enumerate the most important.

Conflicts of Interest

Differentiated units will, inevitably, pursue diverse and often conflicting goals. The result is that for one unit to achieve its goals, it must deprive others of pursuing an alternative set of goals. Such a situation is inherent in differentiation and, hence, a constant source of conflict and, potentially, a cause of disintegration if the conflicts occur among large organizational units and coalitions of these units, or if the conflict occurs among categoric units (e.g., class, ethnic, religious, regional). There can be, then, a zero-sum dynamic built into differentiation.

Moreover, even when units pursue the same goals, the resources available to units may be limited and, as a consequence, competition and Durkheimian selection may ensue. Such competition can move into disintegrative conflict, especially if the actors conflict in a key institutional sphere, if they are large, or if they are the basis for categoric unit formation (as is the case with "religious wars," or tensions among mobilizing classes or ethnic populations). Hence, differentiation creates resource niches which can become disintegrative battlegrounds that polarize significant segments of a population.

Conflict tends to harden categoric boundaries, and in so doing, intensifies conflict—at least in its initial stages. With more clear-cut organization of categoric units in terms of goals, leaders, and symbols, conflict can become less intense as parties negotiate compromises. Hence, if conflicts of interests are well articulated, and accompanied by coherent structures, then conflict need not be disintegrative. But if goals and structures are not clear and if conflicts of interests become viewed in terms of evaluative symbols or value premises, these conflicts will tend to be disintegrative.

Undercoordination

Conflicts aside, coordination of structural units is often problematic, and doubly so if there are conflicts of interests. Regularizing the transactions among units is always difficult, and as differentiation reaches very high levels, the complexity of the system can make coordination extremely tenuous. The result is that conflicts of interests become chronic, or alternatively, units become isolated in localized environments with particularistic systems of symbols.

In either case, there is high disintegrative potential. Chronic conflicts of interest can cause polarization of interests and open conflict; they can generate hostile categoric units with particularistic cultures; and they can force power to be overused, thereby escalating tensions associated with overregulation and inequality. Isolation leads to the development of particularized cultural symbols, categoric unit formation, and interests that will, in the end, come into conflict with other units, setting off conflicts that are not easily resolved because of particularized symbols and categoric definitions of the respective parties to the conflict. However, if categoric units can remain segregated in space, then the particularized symbols of each can work to promote integration. But once contact occurs, especially contact revolving around coordination and distribution of resources, conflicts can become intense, especially if the symbols of each group have developed over a long time and are deeply engrained in their members' emotional and cognitive states.

Overcoordination

Increased power-use (e.g., centralization of administrative/coercive power) or extension of regulatory symbols (e.g., laws and administrative regulations), and typically both, represent responses to the complexity of differentiation and the associated problems of coordination and isolation. Such responses, however, only reduce disintegrative tendencies in the short run, because such levels of control usurp resources, increase inequalities, stagnate the production and distribution on which centers of power depend, engender resentments, and in the end mobilize bases of counterpower (as has been the case with the collapse of the old Soviet Empire).

The balance between over- and undercoordination is difficult to maintain, especially when there is a lack of cultural integration and a state of chronic conflicts of interest. Such is the dilemma, for example, of almost all postrevolutionary systems, which often by necessity must err on the overregulation side which eventually will produce new disintegrative pressures. What is true of postrevolutionary societies is also a point of contention in all societal systems. But if diffuse legitimacy exists, if at least some regulatory symbols are seen as legitimate, and if there are stable media of exchange, especially monetary media, then the capacity to correct for over- or underregulation exists, although only a handful of societies in the last five thousand years has possessed this capacity to a high degree. If the "end history" argument is correct, and all societies move toward legitimated political democracies, perhaps this number will increase, though a cursory glance at the modern world reveals that there is considerable history, much of it revolving around the question of regulation and control, to be played out. A more likely scenario is that the existing and emerging democracies will destabilize with either over- or underregulation.

The Disintegrating Effects of Stratification

Inequality in the distribution of valued resources inevitably generates disintegrative pressures. Marx and Engels ([1848] 1978) as well as Weber ([1922] 1978) were correct in their view that if incumbency in positions across resource hierarchies is highly correlated, if discontinuities between the top and bottom positions in hierarchies are great, if boundaries demarking positions in hierarchies are clear, and if rates of mobility across positions in hierarchies are low, then the potential for mobilization for conflict among those in lower positions is high. More recently, Mary Douglas (1966, 1973) has conceptualized these processes in terms of groups and grids, but the general point is the same: rigid stratification generates categoric units with their own particularistic symbols and sets the stage for conflict.

To phrase the matter somewhat differently, when patterns of resource distribution create clear categoric units of people whose positions in resource hierarchies are, to use Blau's (1977) term, consolidated, formation of conflict groups becomes likely. Yet as Simmel (1956) first emphasized and as later conflict theorists (e.g., Dahrendorf 1959; Coser 1956) stressed, the most violent conflict does not come from highly organized conflict groups, which, in the end, are more likely to have clear goals, leaders, and agendas. As a consequence, such groups are more likely to negotiate and compromise, while acceding to efforts to institutionalize the conflict with regulatory symbols. Such conflicts can, as Coser suggested, have integrative effects.

Stratification becomes disintegrative when it produces violence between (1) a partially mobilized and large categoric unit composed of those in disadvantaged positions and (2) another categoric unit among those in advantaged positions in a system of resource distribution. By partial mobilization, I mean that a well-organized corporate conflict group has yet to form, because (*a*) ideologies and beliefs are not fully articulated or, in fact, are in flux, (*b*) leadership is ambiguous and, typically, unsettled, and (*c*) as a result of (*a*) and (*b*), goals, programs, and agendas are not well formulated. Under these conditions, emotional energies are high and deprivations over inequalities are perceived to escalate; and if a precipitating incident occurs, these energies can lead to conflict, especially if members of a categoric unit are densely settled and are able to communicate their grievances with each other. Such collective outburst can overthrow an existing regime or cause mass repression by centers of power. In both cases, the disintegrative pressures on a system are ratcheted up.

The Disintegrative Effects of Production

Through its effects on increasing the differentiation of corporate and categoric units, to use Hawley's (1986) terms, increasing production can generate disintegrative pressures. Equally important, expanded production initially increases inequalities, as more material surplus is usurped by centers of power and elites (Lenski 1966). As inequalities increase, categoric units become consolidated with organizational units, thereby increasing the discontinuity between the top and bottom segments of social hierarchies and, eventually, raising disintegrative pressures for redistribution of resources. Yet differentiation of production also increases the number and variety of positions, thereby providing more opportunities for mobility and cross-cutting affiliations, which can reduce the disintegrative pressures of differentiation. Still, initial increases in production have, historically, escalated internal conflict up to the very high levels of production that came with industrialization; and even then, other processes in markets and polity can sustain high levels of disintegrative potential.

The Disintegrative Effects of Distribution

Like production, differentiation of distributive processes can increase disintegration, although these are somewhat neutralized by the emergence of money as a legitimating medium, by the networks and cross-cutting affiliations allowed by markets, and by the easier movement of information, people, and materials that comes with expanding capital investments in a distributive infrastructure.

To the degree that money supplants generalized evaluational symbols, however, it can work to increase blatant self-interest and inevitable conflicts of interest among these self-interested actors. Moreover, because money provides elites and centers of power with liquid resources, it escalates usurpation and, hence, inequalities, which then generate consolidated corporate and categoric units with particularistic symbols that can stand in tension or open conflict to each other. Additionally, money allows for the calculation of the value of property, which, in turn, facilitates comparisons among subpopulations over their respective shares of resources; and if such comparisons indicate considerable inequality, they can become the basis for mobilization by categoric units—mobilizations that often increase the disintegrative potential.

Markets also generate great potential for fraud, corruption, exploitation, and abuse because they operate in terms of money and profits, while releasing actors from traditional evaluative and regulatory symbols as well as dense social networks. When such abuses become extreme, resentments accumulate and cause disintegrative pressures to mount. And even if centers of power intervene, such intervention can involve more usurpation of resources, more regulation, and more inequality, all of which increase disintegrative pressures.

Markets' most disintegrative potential, however, resides in their cycles of differentiation, speculation, pyramiding, and collapse. As markets develop, they become highly differentiated and subject to high risk speculation and pyramiding into ever more speculative metamarkets that deal in the terms of exchange in lower markets (Collins 1990). At some point, these markets collapse, and during this phase, the unmet expectations (relative deprivation), the loss of resources for elites and centers of power, and the disruption to networks, routines, and organized interests all increase disintegrative pressures.

The Disintegrative Effects of Power

Although the consolidation and centralization of power can be an integrative force, these processes always contain disintegrative potential. Because I have already discussed these disintegrative potentials in this and in previous chapters, let me briefly review them here.

First, as resources are usurped to sustain centers of power, resentments over such extraction always exist, creating a conflict of interest between these centers and those actors who must give up their resources. Second, as resources are usurped, inequalities increase, thereby escalating conflict-laden potential of all systems of stratification. Third, as centers of power do what they must—that is, regulate and control—they always have difficulty in maintaining balances among the bases of power (incentives, symbols, administration, and coercion). As a consequence, they can fail to sustain legitimacy, use too much coercion, overregulate with administrative and coercive forces, rely too heavily as incentives and underregulate or fail to include important segments of the population in co-optive use of incentives.

And finally, as centers of power engage in geopolitical activities, whether as a response to real or imagined internal and external threats, they inevitably sow the seeds for disintegration, along several fronts: (*a*) the loss of war or failure in the geopolitical economic arena rapidly erodes legitimacy and sets the stage for conflict (even after a series of successes, because one loss can undo all of the legitimating effects of previous successes, for the reasons enumerated below); (*b*) the annexation of new, diverse, and restive populations from success in war increases logistical loads, while establishing new categoric units which can rapidly become conflict-oriented corporate units; (*c*) the mobilization for war requires increased usurpation of resources and, hence, escalated inequality to support administrative/coercive activity that is also used to overregulate actors, with the result that resentments over mounting inequalities or escalated domination ratchet up the disintegrative potential; and (*d*) the extension of spatial territories with success in war or economic co-optation dramatically increases the logistical loads on centers of power for transportation, communication, and administration, thereby hastening the point of overextension that escalates disintegrative pressures.

The Disintegrative Effects of Size

As I emphasized in chapter 2, the size and rate of growth of a population increase logistical loads and, in so doing, pose disintegrative pressures, unless differentiation of productive, distributive, and power structures occurs. Yet such differentiation can, itself, create disintegrative pressures along the lines enumerated above, emphasizing again the fact that population growth sets into motion forces that take on a life of their own and, in so doing, can increase the disintegrative potential of a population.

Growth of a population also tends to increase the density of settlements, and thereby initiate Durkheimian competition for resources and, potentially, Malthusian internal wars that cause the disintegration of a population. The expansion of production and distribution, coupled with the consolidation and centralization of power, can mitigate against this potential by increasing, respectively, the overall level of resources and regulation while, at the same time, generating new resource niches which encourage differentiation of units locked in intense competition. Yet as I have emphasized above, the expansion and differentiation of productive, distributive, and power structures create their own set of disintegrative pressures. Thus, absolute population size relative to space and resources as well as rates of population growth pose constant disintegrative pressures, either directly or indirectly through their effects on differentiation dynamics.

The Disintegrative Effects of Space

The more extensive the territory relative to population size, the greater are the integrative problems. As noted earlier, one set of disintegrative pressures comes from the logistical loads of moving material, people, and information across longer distances. Another set comes from the tendency of subpopulations to become differentiated culturally and organizationally when separated in localized environments.

Under high levels of technological development in communications and transportation, these disintegrative effects are dramatically reduced, although the recent collapse of the Soviet Empire highlights the fact that overregulation of diverse subpopulations in the aftermath of conquest generates a new set of logistical loads that, eventually for all known empires, create disintegrative pressures. Thus, high technology cannot overcome, in the long run, the combined effects of cultural differentiation, overregulation, and inequalities among subpopulations scattered across a large territorial expanse. And once disintegration occurs, it intensifies as cultural differences within and inequalities among subpopulations surface, as has been the case in some subpopulations in the old Soviet bloc or as has persistently been the situation in postcolonial Africa.

In figure 8.3, I have delineated the key structural forces, and their interrelations, that increase disintegrative pressures on a population. Coupled with an inverse of those forces promoting structural and cultural integration, figures 8.1, 8.2 and 8.3 offer a summary view of the ways that human populations, and the structures that organize them, can fall apart.

Population size increases disintegrative potential by directly escalating logistical loads and, indirectly, by selecting for productive, distributive, and power forces that set into motion structural differentiation, political centralization, inequality, and geopolitical processes that create conditions—conflicts of interest, rigid stratification, mounting resentments, and internal threats, conflict-oriented categoric and corporate units, larger territories, and ever more diverse subpopulations—that escalate logistical loads further. As these loads mount, they place direct disintegrative pressure on a population, but they also distort power-use, production, and

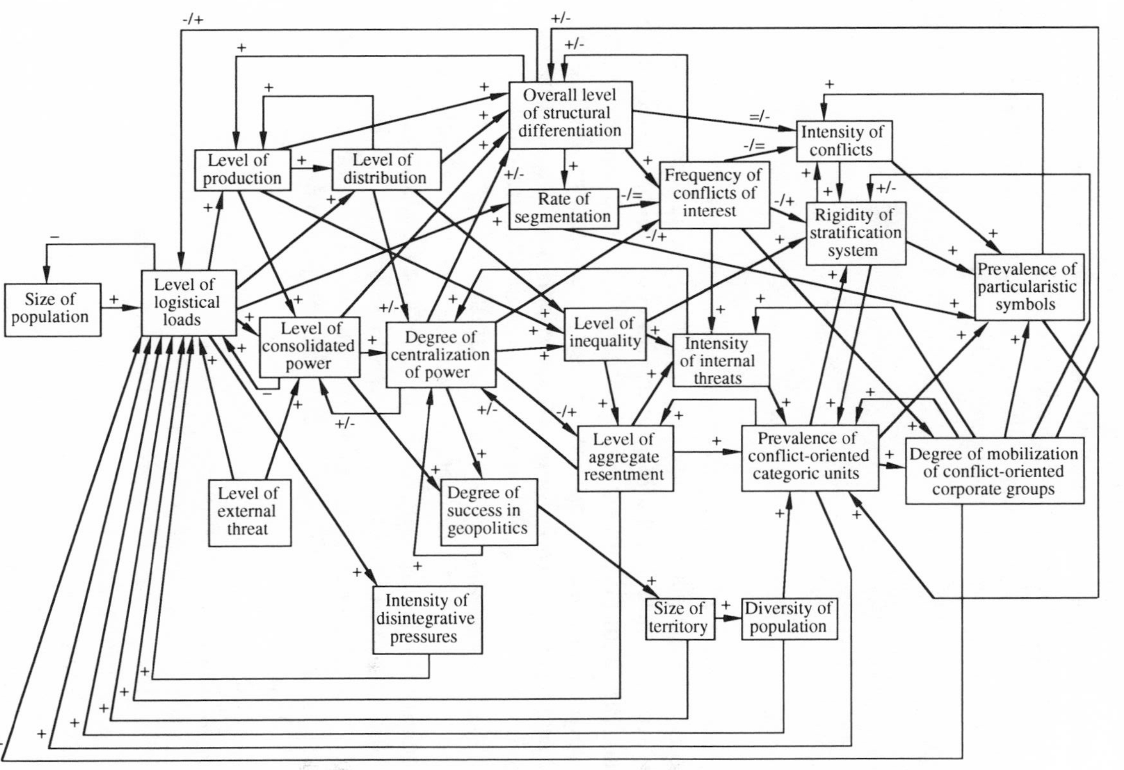

Figure 8.3. A General Model of the Dynamics of Disintegration

distribution in ways that ratchet up inequalities, stratification, resentment, internal threats, conflict categories and groups, and particularistic cultural symbols to new, more-volatile levels, thereby escalating dramatically the disintegrative potential of a population. All, or some combination, of the forces and causal processes outlined in the model have operated for all known societies, and this is why societies disintegrate at some point in their history.

Elementary Principles of Disintegration

This chapter has examined macro phenomena from a number of different angles. As a result, the analysis has become complicated, as I have sought to highlight (*a*) structural and cultural forces as well as (*b*) integrative and disintegrative processes. These are all entangled in a complex web of causal processes, and when we try to picture this web, analysis becomes complex. As in other chapters we can sort through some of the complexity with a list of propositions:

I. The disintegrative potential of a population is:
   A. a positive function of its absolute size and rate of growth relative to its productive, distributive, and regulatory capacities
   B. a positive function of the size of its territory
   C. a positive function of the cultural and organizational diversity among its constituent subpopulations, which, in turn, is a positive function of:
      1. the size of territory
      2. the rate of conquest
      3. the rigidity of stratification
      4. the prevalence of conflict-oriented categoric and corporate groups
   D. a lagged positive function of the degree of centralization of power and regulatory use of power, which, in turn, is:
      1. a positive function of the degree of external threat
      2. a positive function of the level of internal threat, which is a positive and multiplicative function of:

a. the level of inequality, which is a positively curvilinear function of production, distribution, and centralization of power
b. the rigidity of stratification
c. the formation of conflict-oriented categoric and corporate units with particularistic symbols
d. the intensity of conflicts of interest among subpopulations
3. a positively curvilinear function of the level of production and distribution

E. a negatively curvilinear function of the rate of segmentation, which is a positively curvilinear function of a population's size and level of structural differentiation

F. a negatively curvilinear function of the overall level of structural differentiation, which is:
1. a positive function of absolute population size
2. a positive function of production and distribution
3. a positively curvilinear function of the degree of centralization of power and regulation

G. a positive function of the distinctiveness of particularistic symbols of subpopulations, which is:
1. a negative function of the development of generalized symbolic media of exchange, especially money in differentiated market systems
2. a negative function of the degree of correspondence and compatibility among values, beliefs, legitimating symbols, and regulatory symbols
3. a negative function of the degree of structural interdependence, the extensiveness of cross-cutting affiliations, and the rate of positional mobility
4. a positive function of the size of territories
5. a positive function of the degree of segregation of subpopulations in space and in social transactions
6. a positive function of the rigidity of stratification
7. a positive function of prevalence of conflict-oriented categoric and corporate groups

H. a positive function of the intensity of conflict of interest, which is a positive and multiplicative function of:

1. the level of inequality
2. the rigidity of stratification

I. a negative function of the level of structural interdependence among corporate units, which is:
1. a positive function of the scale, scope, volume, and velocity of exchange transactions
2. a positive function of market differentiation
3. a positively curvilinear function of the consolidation and centralization of power
4. a positively curvilinear function of the rate of structural segmentation
5. a positive function of the degree of structural differentiation

J. a negative function of the degree of cross-cutting affiliations, or intersection of parameters

K. a negative function of the rates of positional mobility

Conclusion

This is the last list of propositions for the forces examined in chapters 2 through 8. The analytical models and propositions in these chapters represent my best effort to lay out the dynamic processes involved in developing a theory of macro-level social reality. These chapters are the "toward" part of the subtitle of this book; they point toward a particular set of key forces and their causal relations that should be a part of a theory. None of these forces represents a startlingly new insight; rather, what this book has sought to offer is a theoretical way of (1) reconceptualizing the nature of basic properties of the macro-level universe, though not a dramatic reconceptualization, and (2) visualizing how these forces are interconnected in complex chains of direct, indirect, and reverse causality. They set the stage for the next step in theorizing: to make the theoretical leads enumerated in these pages more parsimonious.

A number of strategies is possible in taking this next step; and as each theorist takes it, we will be closer to a theory of macrodynamics. Initially, I had written a ninth chapter for this book in which I took this next step—showing for all how it should be done. I am more humble now and, hence, have relegated this "grand finale" to an appendix—appendixes almost always being something less than grand. Yet the appendix is an important part of my message: it is how, I believe, we should go about developing theory, because the complex models presented throughout this book are translated and condensed into simple "laws" of macrodynamics. Thus, the long appendix that follows is *my* approach to consolidating into more parsimonious form the theoretical leads developed in chapters 2 to 8. I believe it to be the most appropriate for a science of society—indeed, a social physics. If this assertion peaks interest, please read on.

APPENDIX

Ten Laws of Macrodynamics

Below are my best efforts to use the leads in chapters 2–8 to construct a more parsimonious theory. As I work toward this goal, let me stress several points that have guided my thinking. First, macroanalysis focuses on *whole populations* and their patterns of sociocultural organization in an environment consisting of other organized populations and biophysical parameters. Second, selection processes, both Spencerian and Durkheimian, are key mechanisms by which adaptation of a population to its environment occurs. Third, the organization of a population as a coherent and integrated whole in an environment is only temporary; disintegrative pressures are always operative and eventually overpowering. Fourth, macroanalysis thus consists of explaining the underlying and generic forces that, for a time, enable a population to stave off its eventual disintegration. Fifth, functional, ecological, evolutionary, and conflict theories in sociology present most of the relevant concepts, models, and propositions for developing a theory of macrodynamics. And sixth, and perhaps most importantly, certain propensities of social organization are "driving forces" in that the operation of other forces organizing a population is most affected by the values of, and interaction effects among, these more primary forces. These driving forces are (1) population size, (2) production, (3) distribution, and (4) power. If we know the values of these forces as they intersect with each other, we will know a good deal about the operation of all other forces organizing a population, as well as the reverse causal affects of other forces. Thus, my general theory of macrodynamics begins by articulating some elementary principles on these four basic forces.

Law 1: Population

The size and rate of growth of a population are, in my view, the most central demographic forces in the organization of humans. For ultimately, as emphasized in chapter 2, the size of a population—both in the long-term evolution of the species and in any particular society today—sets into motion other macrodynamic processes by increasing logistical loads for maintenance, coordination, and control as well as by intensifying Malthusian pressures to the point of a dramatic correction through war (both external and internal), disease, pestilence, and famine. Of course, once in motion, the driving effects of population recede, unless overpopulation dramatically raises logistical loads.

The forces causing increases in population size have, however, been somewhat undertheorized because its values are greatly affected by history, local conditions, and periodic events. Nonetheless, because population size is so central to macrodynamic processes, we should be able to denote those basic forces affecting its variation, even after the specifics of history, contingency, and unanticipated events are acknowledged. Below is an elementary principle on population size:

$$N = (+/-P) \times (DF) \times (PO_{CL}) + (TS) \qquad (1)$$

where:

N = the size of a population
P = the level of production
DF = the level of differentiation among corporate units, categoric units, and systems of symbols
PO_{CL} = the degree of political (PO) consolidation ($_{CL}$)
TS = the amount of territorial space inhabited and controlled by a population

Before discussing equation (1) and the nine other equations to follow, let me offer a word on the unconventional notation. My goal here is not to be a "mathematical sociologist," but merely to state certain basic relationships among the forces specified in the theory. These quasi-equations could, of course, be put in standard notation, but my experience is that this format intimidates the nonmathematically inclined, including me. Instead, I have found this simplified format communicates best to a wider audience—once they get over the horror of "math."

The principle represented in equation (1) states that the size of a population (N) is a positive curvilinear function of the level of production (P), while being a positive function of the degree of political consolidation of the four bases of power (PO_{CL}), the overall level of sociocultural differentiation (DF), and the size of the territory occupied by a population (TS). As the multiplication signs among (P), (PO_{CL}), and (DF) indicate, these influence the values for (N) by virtue of their effects on each other. The proposition states, in essence, that if one wants to explain the size of a population, the forces listed on the right side of the equation are the most generic and important. The processes by which each of these forces has exerted its influence on the size of a population have been delineated in the various models presented in chapters 2 to 8. Yet these causal paths can be simplified considerably when our attention is focused by an equation. Now, as I will for each equation to be presented, let me briefly discuss the relationship between the focal force—in this case (N)—and each of those forces on the right side of the equation that determines its values.

Production and population size

As emphasized in chapter 2, the size of a population is ultimately related to the capacity to produce sufficient goods to sustain its members. If increases in production occur, a greater quantity of goods and commodities exists which can then be used to increase the capacity to sustain a larger population. At the same time that this sequence of events occurs, increasing production of goods and commodities is also raising the normative standard of living which operates, in the long run, to ratchet up the requirements for what is defined as minimal sustenance, thereby reducing the capacity for supporting a population and eventually causing the stabilization of, or actual decline in, the size of a population (i.e., the end phase of the "demographic transition"). Thus, increasing production bears a complicated relationship to population size because it enables a population to grow but creates systems of cultural symbols that operate to slow this growth (as is denoted by the +/− sign signaling a positively curvilinear relationship).

Power and Population

The consolidation of the four bases of power (i.e., coercion, administration, legitimation of symbols, and material incentives) initiates the centralization of power (or CO_{CT}) and the combination of consolidation and moderate centralization increases the capacity to coordinate and control a larger population. I have emphasized CO_{CL} in equation (1) because I see this as the critical breakthrough for a growing population, with a moderate degree of (PO_{CT}) being a by-product of (PO_{CL}), as will be explored in equation (5). Indeed, as I emphasized in chapter 5, a high degree of (PO_{CT}) undermines balances achieved with (PO_{CL}) and biases power toward coercion and overregulation by its administrative component. And if these balances are undone, new Malthusian pressures toward internal and external war mount and new kinds of logistical loads for coordination and control escalate. If the Spencerian selection pressures thereby generated go unmet, and this becomes increasingly the case when (PO_{CL}) is undermined by (PO_{CT}), then disintegration of a population and decreases in its size may ensue, as I will argue in equation (10). But, the key argument in equation (1) is that consolidation sets into motion capacities for regulation (via moderate centralization) that enable a population to grow because of increased capacity for coordination and control.

Differentiation and population

Herbert Spencer was the first to theorize that differentiation (DF) was essential to a growing "social mass." For as the "mass" grows, the complexity of structural supports necessary to sustain this mass also multiply; and such differentiation proceeds, a population can continue to grow. There is a certain vagueness in Spencer's approach (see fig. 2.3), but the general idea is sound, although it requires amplification from bioecological theory.

As differentiation increases, the number of diverse resource niches that can support organizational units, social categories, and systems of symbols also increases. Such niches generate opportunities for additional structural units, categories, and symbol systems, which, in turn, can be used to sustain and organize members of a population. Thus, one level of differentiation in structural units, categories, and symbols creates the resource niches that beget further differentiation. The result is an increased capacity to support and organize a larger population.

Space and population

It is obvious, of course, that a larger territory can, other things being equal, sustain a bigger population than a small territory. But this relationship is, nonetheless, fundamental: the size of a population is, eventually, constrained by the amount of territory it controls and by the resource levels within this territory.

As the size of territories increases, there is more living space, unless it is uninhabitable for geographical reasons. Whether habitable or not, a larger territory will generally increase not only the total amount of living space but also the total level and diversity of resources potentially available for productive activity. In turn, all of these forces increase the capacity to support a larger population.

Law 2: Production

The capacities to extract resources from the environment, convert them into usable goods and commodities, and distribute products and related services set parameters for other driving forces. For production and distribution determine how many members of a population can be supported as well as how much material surplus is available to support centers of power. Expanding production and distribution lower both the logistical loads and Malthusian pressures on a population by increasing the number who can be sustained and by enabling coordination and control of members through the consolidation and centralization of power.

As I argued in chapter 4, distribution dynamics should be separated from production processes, because once exchange distribution reaches certain levels of complexity, revolving around differentiated market systems and use of money and other generalized media, it has independent effects on all other macrodynamics, including production. Indeed, a high level of market differentiation acts as a supercharger on virtually every facet of human organization. Thus, as I did with chapters 3 and 4, I will separate the analysis of production and distribution in order to develop an elementary principle on each. Let me turn to production first.

The basic question thus becomes: What forces affect the level of production? Again, a simple equation can provide an answer at a very general theoretical level.

$$(P) = [(TE) \times (CA_{PH}) \times (CA_{HU}) \times (DF_{MS,MY}) \times (NR)] + N \qquad (2)$$

where:

TE = the level of technology for manipulating both the biophysical and social environment

CA_{PH} = the level of physical ($_{PH}$) capital (CA) formation, including liquid capital that can purchase physical capital

CA_{HU} = the level of human ($_{HU}$) capital (CA), especially knowledge and skill levels of individuals

$DF_{MS,MY}$ = the level of differentiation (DF) in market systems ($_{MS}$) and generalized money media ($_{MY}$)

NR = the level of access to natural resources

N = the size of a population

To phrase the equation in words, the level of production is a positive function of (*a*) the level of technology (TE), broadly defined as the capacity to manipulate all aspects of the physical, biological, and social environment, (*b*) the level of access to natural resources, whether a population's own or those of another population (NR), (*c*) the level of physical capital (CA_{PH}) that can be devoted to productive processes, (*d*) the level of human capital (CA_{HU}), or knowledge and skill among those actors involved in productive processes, (*e*) the level of differentiation of markets and extensiveness of money and related generalized media in distributive processes, and (*f*) the size of a population. Forces (*a*) through (*e*) are multiplicatively related to each other in their effects on production. In contrast, population size exerts an independent effect on production (and vice versa, as is emphasized in equation [1]). As I did for equation (1), let me now delineate some of the causal processes connecting forces denoted on the right side of the equation to production.

Technology and Production

Technology affects production by virtue of its consequences for creating: knowledge about varieties of resources in the biophysical environment and how to gain access to such resources; knowledge about how to construct and use physical implements and energy to gain access to, as well as to convert and distribute, resources; and knowledge about entrepreneurial mechanisms for organizing capital and energy in conjunction with human labor to gather, produce, and distribute resources. Thus, as technology increases, these bases of knowledge grow and, as a consequence, production can potentially increase.

As production increases, new technologies are acquired and accumulated, and if the institutionalization of knowledge-producing activity also occurs (i.e., the creation of specific sociocultural systems for science and engineering), then rates of acquisition and accumulation are dramatically accelerated, thereby raising the level of technology and, hence, the potential for growth in production. Such potential depends upon market differentiation to encourage the distribution of accumulated knowledge to ever more productive units and power-use to facilitate the institutionalization of science and engineering in ways that attach science and engineering to market dynamics (Historically, this first involved status group competition among elites, but once technology became practical for war and production, the broader institutionalization of science and engineering was initiated).

Physical capital and production

For production to reach high levels, it is obvious that there must be a physical capacity to build machines harnessed to inanimate sources of energy, construct buildings capable of housing machines and the labor attending them, install an efficient communication and transportation system, and erect facilities for housing service and administrative systems. Such capacities are dependent upon technological development, as well as upon the pool of liquid capital to finance their construction and operation, with this pool being related not only to production and distribution but also to the taxation-spending policies of centers of power.

Human Capital and Production

The knowledge and skill levels of those engaged in productive activity greatly affect the efficiency and scale of such production. Human capital is both a cause and a by-product of increased production, both directly and indirectly. Directly, expanding production differentiates and organizes labor in ways that generate demands in a labor market for diverse skills. The consolidation of power, which is made possible by the surplus caused by expanding production, operates indirectly to increase knowledge and skill levels through the extension of education (often for purely political indoctrination but also for economic development) as well as through the institutionalization of science, especially when this institutionalization is lodged in "higher" educational structures or "institutes" of science. The growth of centers of power, the institutionalization of science, and the extension of education themselves expand labor market demands for knowledge and skill. And once expanding levels of production, government, education, and science exert these effects on labor markets, market processes constantly ratchet up the level of human capital. In fact, credential inflation can occur, as access to positions in economic (science, education, and power) structures is defined in terms of "necessary" credentials (that presumably "certify" knowledge and skill). When credential inflation ensues, the level of human capital can expand beyond the capacity of the labor market to secure positions in productive processes. But, despite such inflation of credentials, and the consequences of this for wages and governmental policies, the availability of an educated labor pool works to increase the level of production by providing a sufficient number of workers with necessary skills and knowledge.

Market Differentiation and Production

At the center of expanding production are market processes. Of course, increasing levels of production create Spencerian selection pressures for the differentiation of markets as well as for the use of money, credit, and related financial instruments, but once differentiated markets employing money (and related media of exchange) exist, they can operate independently of gathering and conversion processes, eventually to the point of pyramiding, overspeculation and collapse. Even as they do so, however, they increase demands for productive outputs for not only goods but increasingly services. As this demand escalates, new resource niches are generated and new kinds of productive units emerge in these niches, up to the point where organizational density sets into motion Durkheimian competition among productive units. Such competition, however, is not purely Darwinian because less "fit" organizational units can do more than move to a new niche or die; they also create new resource niches by producing goods and services that not only respond to market demands, but also create such demands.

Thus, competition among productive units in niches often works to expand and differentiate markets in ways that escalate demands for expanded productive activity. Markets thus provide not only the resources for productive units by virtue of translating needs into demands, they are also the vehicle by which new productive outputs and combinations create demands that, in turn, increase the number of resource niches for productive units. Advertising is, of course, the culmination of these dynamics in capitalist systems. Indeed, advertising can create the illusion of new products when in fact little is new, but this does not negate the fact that new markets are established and, hence, more resources are available for productive units. Market systems without this capacity for hyper-differentiation work to stagnate demand and, hence, production at relatively stable levels of output. Moreover, they cannot generate the aggregate wealth—and, of course, neither the chains of overspeculation that lead to collapse nor the hording that increases inequalities and conflict—that stimulates production.

Natural Resources and Production

Access to natural resources necessary for production is, to a great extent, a reflection of (*a*) the level of technology or knowledge about what resources are usable and how they are to be gathered, (*b*) the level of existing production that has the organizational capacity to secure and use resources, (*c*) the level of market and infrastructural (physical capital) development to stimulate searches for resources, and (*d*) the level of power that can be mobilized to facilitate (*a*), (*b*), and (*c*) above and to engage in geopolitical activity to secure resources outside a population's territory.

This last force becomes particularly important when a population's territory is resource-poor, and as a consequence, it sets into motion patterns of intersocietal war and trade. If intersocietal markets are not well developed or if they are controlled by one society or distributive cartels, then intersocietal war becomes more likely as a means to secure resources; and once war becomes defined as competition for resources, it takes on a chronic character, up to the point of temporary hegemony of one state or exhaustion and collapse of states.

In the latter case, market forces can evolve in the coercive vacuum, but even when a temporary hegemony or a successful cartel exists, the sheer movement of goods from one territory to another creates the physical infrastructure and the market forces (suppliers, dealers, and consumers) that, in the long run, can lead to less regulated trade. When fewer restrictions exist, more resources can be secured at lower costs (because excess profiteering by cartels is reduced and/or the costs of maintaining a coercive-administrative structure are also reduced); and as a consequence, production is directly stimulated by allowing access to resources and, indirectly, by creating new sources of demand beyond a population's border for the outputs of production.

Thus, the level of natural resources involves much more than what is contained in a territory. Actual *access to* resources is the more important dynamic, as influenced by technology, production, markets, geopolitics, and patterns of intersocietal trade.

Population and Production

The size of a population affects production in several basic ways: (*a*) as a source of Malthusian pressure and Spencerian selection for new productive outputs when growth exceeds current productive capacities; (*b*) as a source of Spencerian selection for the consolidation of power, which, in turn, creates demands for, and facilitates the development of, the surplus productive outputs on which the consolidation of power depends; (*c*) as a source of potential market demand, which, if a population is large, differentiated, and exhibits high normative standards of living can stimulate production and which, if a population is small, can limit production because of the inability to generate a large aggregate or differentiated demand, although a high normative standard can increase the level of demand in a smaller population for productive outputs; and (*d*) as a source of productive stagnation when a population becomes so large as to consume the material surplus of productive and political units in ways that inhibit capital formation and technological development.

Depending on which profile of (*a*), (*b*), (*c*), and (*d*) is evident, production will vary. For example, if (*a*) prevails at the same time as (*d*), then innovations and their implementation are less likely, creating a potential for not just stagnation but a Malthusian correction. If (*a*) and (*b*) exist simultaneously, then the implementation of innovations becomes more likely, unless power has become overly centralized around an administrative-coercive profile. If (*c*) prevails for a population with a high level of aggregate demand, as well as a high level of diversity in demand, then productive-outputs are likely to increase, unless condition (*d*) also prevails. If (*c*) exists for a small population, then productive outputs will be export oriented, especially when condition (*b*) also exists. Other profiles and scenarios are possible, but the critical point is that population size exerts enormous influence on production.

Law 3: Distribution

In chapter 4, I divided distribution processes into "infrastructural" and "exchange." Here, I will concentrate on exchange, because the discussion of physical capital above (CA_{PH}) incorporates most of the ideas examined in chapter 4 and places them in the proper context—that is, as a crucial condition for increased production and, as we will see, market differentiation. As I phrased the question in chapter 4 and in equation (1), it is the differentiation (DF) of market systems ($_{MS}$) and the use of money and equivalent media ($_{MY}$) that increase the exchange of goods and services. Thus, the elementary principle can be phrased in terms of ($DF_{MS,MY}$):

$$DF_{MS,MY} = (P) \times (PO_{CL}) \times (+/-PO_{CT}) \times (+/-N) \times (CA_{PH,MY}) \qquad (3)$$

where:

P = the level of production
PO_{CL} = the degree of political (PO) consolidation ($_{CL}$)
PO_{CT} = the degree to which power (PO) is centralized ($_{CT}$)
N = the size of a population
$CA_{PH,MY}$ = the level of physical ($_{PH}$) capital (CA) formation, including liquid capital ($_{MY}$) that can purchase physical capital

To rephrase in words, equation (3) simply states that the level of differentiation of markets (into diverse types, hierarchies, and meta-systems) and the use of money (and its equivalents) are the result of the multiplicative relations among production (P), the consolidation of power (PO_{CL}), the centralization of power (PO_{CT}) (which, after a certain threshold has negative effects on [$DF_{MS,MY}$]), the size of a population (which, also, has negative effects on [$DF_{MS,MY}$] when too large), and the level of physical capital formation ($CA_{PH,MY}$). Each exerts an independent effect on ($DF_{MS,MY}$), but the interaction of these forces on each other generates an effect on ($DF_{MS,MY}$) that far exceeds their additive effects alone.

Production and market differentiation

As already emphasized in equation (2), production and markets are intimately connected. Emphasizing here only the effect of production on markets, increased production generates Spencerian selection for expanded distribution using ever more neutral media of exchange. Market differentiation can occur somewhat independently of productive outputs of goods and commodities, but without a base of exchange transactions involving "hard" goods and "necessary" services (i.e., unrelated to speculative activities), such differentiation will involve considerable speculation in the media of exchange and, as a consequence, be subject to crisis and collapse. Hence, the greater the ratio of "hard" and "necessary" productive outputs relative to outputs that service speculation in metamarkets, the more stable will be patterns of market differentiation and, hence, the less susceptible will markets be to overspeculation and collapse.

As I noted for equation (2), production is often intentionally geared to creating new levels of market demand and differentiation. Such is particularly likely to be the case under conditions of intense competition among productive units who, facing falling rates of profit, seek to create "new markets" (i.e., new resource niches) for existing products and new products for new market niches. Such effects lead to rapid differentiation of markets, as targeted consumers are enticed to exchange for new productive outputs. And once this connection among competition, falling profits, new outputs, and market differentiation reaches a certain threshold, it becomes self-sustaining and self-escalating up to the point where aggregate levels of income and wealth which can be used for consumption are exhausted. For as "niche marketing" is established as a viable strategy of productive units, it encourages producers to create new output combinations, and it establishes a market system built around ever more specialized exchanges.

Consolidation of power and market differentiation

The differentiation of markets is limited without the consolidation of the four bases of power. Indeed, if markets develop and differentiate in the absence of power, the latter is soon mobilized under Spencerian selection pressures (*a*) to assure that agreements are honored, (*b*) to prevent external coercive or mercantile forces from intervening, (*c*) to coin as well as protect the value of money and other media, (*d*) to maintain cartels and monopolies, and (*e*) to build and maintain transportation and communication infrastructures necessary for exchange transactions.

Depending on *how* power is consolidated, the profiles among (*a*) through (*e*) above will vary. If coercive and administrative bases dominate over legitimating symbols and material incentives, then market differentiation will be limited and less dynamic because noncompetitive cartels and monopolies will be maintained, speculation over money and other media will be restricted in order to maintain control of productive activity, and coercive and administrative protection of internal markets or restricted intersocietal markets will be extensive. If, on the other hand, use of material incentives, reliance on legitimating symbols, and moderate use of administrative capacities dominate over coercion, then market differentiation will accelerate because money and other generalized media will be viewed as a source of legitimation, speculation and investment will be seen as a source of surplus wealth to support centers of power, use of material incentives to encourage new productive outputs for achieving political goals will be seen as critical to political legitimacy, and administered competition within internal and intersocietal markets will be viewed as a more efficient and less costly way to realize political goals than coercive protectionism.

Centralization of Power and Market Differentiation

Some degree of centralization of power comes with consolidation of the four bases of power, and without this centralization market differentiation is limited because the forces that make markets dynamic—stable money and other generalized media, enforcement of contracts, protection from predatory practices, managed competition, and incentives for new productive outputs—cannot be fully realized. Yet if power becomes highly centralized, using a combination of administration and coercion to regulate markets and to extract material surplus for maintaining the administrative-coercive apparatus, then such a system creates disincentives for market differentiation, along several fronts: profit incentives for productive innovation and creation of new market niches decline; speculation with, and over, money and other generalized media is restricted; material incentives for new productive outputs and market niches are dominated by concerns of administrative control; and surplus wealth and potential capital investment are extracted at high rates in order to sustain administrative-coercive control.

It is difficult to sustain a balance between underregulation, where hyper-differentiation and metamarket speculation eventually produce market collapse and de-differentiation, and overregulation, where administrative-coercive control limits speculation, investment, and innovation to the point of stagnation. Consolidated power creates pressures for more regulation, pressures which are intensified when markets overdifferentiate because of speculation; and yet overregulation of markets and speculation cause productive stagnation and market stasis, thereby intensifying pressures for less regulation. Few political systems have maintained a balance between these extremes for more than a few hundred years.

Population and Market Differentiation

The differentiation of markets is ultimately limited by the size of a population, although the aggregate wealth, coupled with the normative standards of living and consumption among members in a population qualify the effects of population size on market differentiation. But ultimately, a small population, even if wealthy and oriented for consumption, cannot sustain a highly differentiated market system within its territories; it must ultimately rely upon the production facilities and market systems of other societies. A large population can, potentially, support a more differentiated market system, *if* production, wealth, and consumption standards are high. Obviously, many large populations of the past and the present do not evidence large-scale or highly differentiated markets because production is inadequate, political overregulation creates disincentives for innovation and investment, exploitive extractive practices of other societies have inhibited production, and overpopulation saps productive surplus and makes political stability problematic. Yet such societies can potentially reveal large and differentiated markets in ways that small societies, even when wealthy, cannot. Hence, there is a rough positively curvilinear relationship between population size and market differentiation: market differentiation will increase with size, up to the point of overpopulation saps all surplus and prevents centers of power from achieving the stability necessary to facilitate production and market development.

Physical Capital and Market Differentiation

The level of physical capital, as it affects production and consolidation of power, determines indirectly how differentiated markets can become.

More directly, the amount of liquid capital and the extensiveness of transportation and communication infrastructures directly affect market differentiation. Liquid capital can be used flexibly to construct physical infrastructures for administration of exchanges, movements of goods and services, and flows of information in ways that facilitate market differentiation. Without liquid capital, the fixed infrastructure places limits on market differentiation because existing systems of ports, roads, exchange locations, information channels, and other physical facilities dictate where and how exchange transactions can occur. With liquid capital, physical facilities can be altered, moved, reconfigured, and expanded in ways providing the necessary infrastructural base for expanding the volume, velocity, and diversity of exchange transactions.

Thus, more than fixed physical structures are essential to market differentiation; liquid capital surplus must also exist to provide for further development and alteration of the infrastructure. Such capital emerges when productive units, markets, and centers of power create and invest surplus in new infrastructures. If this is not possible, then market differentiation will be inhibited.

Laws 4 and 5: Power

As I stressed in chapter 5, two dimensions of power are important for macro structural analysis: (1) the consolidation of power (PO_{CL}), or bringing together coercion, administration, incentives, and legitimation; and (2) the centralization of power (PO_{CT}), or concentrating the capacities and facilities for decision making, coercion, and administration. These two can vary somewhat independently of each other, although consolidation of power is an initial cause of centralization, whereas centralization can undo balances in consolidated power. Thus, for power, we need at least two elementary principles, one on consolidation (PO_{CL}) and another on centralization (PO_{CT}). Let me turn first to consolidation.

$$PO_{CL} = (N) + [(P) \times (DF_{MS,MY})] \tag{4}$$

where:

PO_{CL} = the degree of political (PO) consolidation ($_{CL}$)

N = the size of a population

P = the level of production
$DF_{MS,MY}$ = the level of differentiation (DF) in market systems ($_{MS}$) and generalized money media ($_{MY}$)

In more discursive terms, this proposition states that the consolidation of the four bases of power (material incentives, legitimating symbols, coercion, and administration), or (PO_{CL}), is a positively curvilinear function of absolute size of a population, plus the combined multiplicative effects of market system differentiation in terms of money, or ($DF_{MS,MY}$).

Population Size and the Consolidation of Power

As emphasized in chapter 2, a growing population generates intense Spencerian selection pressures for coordination, control, and regulation through the mobilization of power. Such mobilization involves creating and organizing an administrative structure, a capacity for coercion, an ability to control and manipulate material incentives, and a capacity to produce legitimating symbols. Unless this consolidation occurs, disintegration is eminent. This relationship between population size and consolidation is often positively curvilinear, increasing consolidation as a population grows but undermining consolidation when populations get very large and set into motion uncontrollable disintegrative pressures (see equation [10]).

Production and the Consolidation of Power

The level of production, as fueled by the differentiation of market systems (DF_{MS}) and the availability of liquid or money capital (CA_{MY}), exerts a direct influence on the consolidation of power by creating the material surplus that is usurped by centers of power to support and sustain administrative structures, coercive forces, and pools of material incentives. Without this material surplus, the scale and scope of political consolidation are limited, but once production increases and provides a material surplus, consolidation occurs very rapidly, especially under the Spencerian selection pressures for coordination and control that a growing population generates.

Markets and the Consolidation of Power

Market differentiation ultimately causes the development of money and other liquid media of exchange, which, indirectly, influence political consolidation through their effects on production. More directly, markets facilitate the consolidation of power by creating additional wealth which can be usurped and used to sustain administrative, coercive, and incentive systems. Equally important, the existence of money becomes a powerful, secular source of legitimation by virtue of its capacity to bestow and create value for members of a society. Indeed, consolidation of power is somewhat limited when legitimating symbols are tied to religious beliefs, for under this condition coercive-administrative structures will dominate incentive systems, which must be constrained and circumscribed by conformity to religious dictates. Some societies, such as contemporary Iran, have been able to sustain both a monetary and religious system of legitimation but the secular propensities of money and markets stand in tension with the sacred dictates involved in the religious legitimation of power whose activities are supported by money secured from international market activity. It is perhaps the reliance on external markets for money rather than internal market differentiation that enables this system of consolidated power to endure.

But more typically, market differentiation and the use of money shift the bases of legitimation to more secular sources, such as law and generalized media of exchange. At the same time, markets using money provide the liquid capital necessary to finance administration, coercion, and incentives. For until markets generate liquid wealth, the consolidation of power is limited by the lack of financial capacity to sustain independent administrative, coercive, and incentive systems based upon secular legitimating symbols.

Having reviewed the terms in equation (4) on the consolidation of power, let me now turn to the centralization of power, or (PO_{CT}):

$$PO_{CT} = (+/{=}PO_{CL}) + (I) \times (TH_{IN}) \times (TH_{EX}) \qquad (5)$$

where:

PO_{CT} = the degree to which power is centralized

PO_{CL} = the degree of political (PO) consolidation ($_{CL}$)

I = the degree of inequality in the distribution of valued resources among members of a population

TH_{IN} = the level of internal ($_{IN}$) threat (TH) stemming from actual conflict, or perceived potential for conflict among organizational units and subpopulations

TH_{EX} = the level of external ($_{EX}$) threat (TH) stemming from actual conflict with other populations, or perceived potential for such conflict

Stated more discursively, equation (5) argues that the level of (PO_{CT}), or the degree to which coercive capacities, decision making, administration, and control of material incentives are controlled by a limited number of individuals and organizational units, is positively related to the initial consolidation of power (PO_{CL}) and the multiplicative relationship among the level of inequality (I), internal threat (TH_{IN}), and external threat (TH_{EX}).

Consolidation and the Centralization of Power

As emphasized in chapter 5 and elsewhere above, the centralization of power is initiated with consolidation. For, in pulling administration and coercion together, some centralization of decision making, administration, and enforcement must occur. Having to balance these events with needs for legitimating symbols and co-optive use of material incentives can mitigate against further centralization of power. Yet once an administrative-coercive profile exists, centralization is difficult to stop, even at the cost of losing some legitimating symbols and abandoning co-optive use of material incentives. Thus, the principal effect of consolidation is during the early stages of centralization. Once coercive-administrative structures exist, further centralization is difficult to stop, and so, centralization begets more centralization, unless other forces intervene.

Inequality and the Centralization of Power

The centralization of power is, itself, a contribution to inequality because, along with material well-being and prestige, power is a highly valued resource. As power is concentrated, it is used to acquire additional shares of not only power but material wealth and prestige as well. But more is involved, because the centralization of power not only causes inequality, it is dependent upon such inequality. Power is primarily sustained by the capacity to usurp surplus material wealth and, secondarily, to imbue leaders with legitimating prestige. If there were no inequalities in material wealth and prestige, then sufficient quantities of these resources would be difficult to extract. Thus, the dynamics of power require available pools and concentrations of resources to tax and usurp; and although taxes on all members of a population can generate resources, it is only when pools of surplus wealth exist that effective usurpation can occur. In highly productive populations, such surplus exists at the household level and can be taxed at a graduated manner, whereas in less productive populations, it is pooled in the hands of local and regional elites, and then taxed by centralized authority. In either case, inequalities in income and wealth distributions facilitate usurpation.

Inequality also influences centralization of power by virtue of its effects on stratification, which, in turn, escalates internal threats from incumbents in various strata or classes. The greater the inequality in resource distribution, the more correlated these distributions, the more discontinuous points of similar resource shares along these distributions, and the less mobility across these points, then the more stratified is a population and the greater is the potential for conflict. Such potential, or actual conflict, creates internal threats, which cause the centralization of power, especially its coercive base to deal with such threats.

Internal Threats and the Centralization of Power

Although stratification processes stemming from inequalities in resource distribution are the major impetus to internal threat, there are other sources, such as (*a*) conflicts of interests among organizational units (economic, religious, political, kinship) that can erupt into actual conflict and (*b*) tensions among categoric units (regional, ethnic, religious, and subcultural) that can also evolve into conflict. Typically, inequalities are associated with these sources of conflict, but such need not always be the case. Whether correlated with class divisions or not, these forms of conflict force the centralization of power, especially its coercive-administrative bases. Such centralization and use of coercive power can, in turn, undermine political legitimacy, thereby requiring centers of power to rely on ever more coercion and administrative monitoring in the absence of accepted legitimating symbols.

Such a situation can spiral to the point where centers of power cannot control internal conflict, with the result that the population disintegrates for a time. In the aftermath of disintegration, new centers of power will have to rely heavily on coercion to deal with persisting threats. And so, even if internal threats cause political collapse and disintegrative chaos, the aftermath of internal conflict often involves the imposition of a new, highly centralized political regime.

External Threat and the Centralization of Power

When a population experiences threats from outside its borders, power is mobilized and centralized to cope with this threat—whether it be military, economic, cultural (e.g., unwanted religious infiltrations) or demographic (e.g., undesired immigration). Military threat is the most likely to prompt centralization of coercive power, whereas economic, cultural, or demographic threats will centralize administrative structures and, if necessary, coercive capacities as well.

It is not essential that the threats be real, only that they are perceived to exist. Indeed, as noted in chapter 5, centers of power often manufacture external threats to bolster a shaky political regime. There may also be a curvilinear feature to external threats: low levels of threat have little effect because the threat is so low as to invite indifference; moderate levels centralize power immediately because the threat is seen as real and manageable; and very high levels of threat may be so great as to create a sense of futility that translates into resistance to, and defections from, efforts by centers of power to centralize further.

Equations (1)–(5) are, in essence, relatively simple propositions on those forces that drive macrodynamics. They are "driving forces" because everything else that occurs at the macro level of human organization is dictated by the values for, and interactions among, these forces. Of course, the values of these driving forces are greatly influenced by other subforces of human organization—that is, inequality, external and internal threats, human and physical capital, technology, resource levels, ecological space, and differentiation—but the key set of forces behind each of the driving forces is the remaining driving forces. Moreover, if an equation is constructured for each of the subforces that are also involved, these too would be shaped primarily by the driving forces.

In making these assertions, I am not deviating far from existing theories which tend to emphasize production and power, although I have perhaps emphasized market differentiation and population size more than most macro-level theories. And I have probably placed less emphasis on inequality, stratification, and internal conflict than many conflict-oriented theories because I tend to see these as derivatives of the driving forces, though they have reverse causal effects on power dynamics.

If one counts all of the driving forces and subforces in equations (1)–(5), there are not very many. And even if we add several more to a list of these basic forces, there are still not very many. The reason for this is that the complexity of the social universe comes from the intersections and interactions of forces rather than from the actual number of such forces. I will add only five more equations, indicating that macrodynamic processes can, at a most basic and abstract level, be understood in terms of less than ten elementary principles. I could add a few more principles, if I wished to construct a separate proposition for each term in all of the equations, but even if I performed this exercise, the number of elementary principles is relatively very small, because the same sets of forces reappear in varying combinations in all of the principles.

What additional propositions are essential, then, to complete my list of elementary principles of macrodynamics? From the perspective developed in chapters 6, 7, and 8, three additional sets are necessary for (1) spatial dynamics, (2) differentiation dynamics, and (3) disintegrating dynamics.

Law 6: Territory

The amount of territory and the patterns of settlement in a territory are two distinguishing features in the organization of a population. As I noted in chapter 6, the overall size of territory and settlement patterns within territories are often conceptualized in somewhat different theoretical literatures—geopolitics and urban ecology, respectively. The composite model drawn in figure 6.5 attempted to show the interconnection of these two dimensions of spatial organization, although they operate in terms of somewhat different dynamics. For purposes here in developing elementary principles, I will present two principles to highlight these varying dynamics. One will be on the amount of territorial space inhabited and controlled by a population, or (TS), and the other will be on the density of settlement patterns within a territory, or (TS_D).

In equation (6), the forces affecting the size of a population's territory are enumerated. To a degree, the size of territory is historically contingent, and yet the historical events leading to the occupation and control of a territory have proceeded in terms of the forces enumerated in equation (5).

$$TS = (N) \times (P) \times (PO_{CL,CT}) \times EC_S \qquad (6)$$

where:

TS = the amount of territorial space inhabited and controlled by a population

N = the size of a population

P = the level of production

$PO_{CL,CT}$ = the level of consolidated ($_{CL}$) power (PO) and the degree of centralization ($_{CT}$) of such consolidated power

EC_S = the degree of success ($_S$) in external conflict (EC)

To translate into words, the size of the territory inhabited and controlled by a population is a positive function of the multiplicative effects among population size (N), production (P), consolidation of power (PO_{CL}), centralization of power (PO_{CT}), and success ($_S$) in external conflict (EC_S). Let me briefly examine each of these relations.

Population Size and Space

A large population will require more territory, once it surpasses the capacity to absorb its members in dense settlements. Thus, population size, per se, exerts considerable effort on the amount of space a population will inhabit and control. But much of the effect of population size is indirect, exerting itself through its consequences for production as the latter affects power.

Production and Space

As with population size, production exerts its influence on the amount of territory controlled by a population through its effects on population size and power as these determine the outcome of external conflict. But there are more direct effects: expanding production causes the development of new transportation and communication technologies as well as the application of these technologies into physical capital that supports a distributive infrastructure that, in turn, facilitates and encourages movement in space and, eventually, movement out into new territories; and production also escalates distributive processes leading to market differentiation that, ultimately, facilitates the movement of goods and people beyond territorial borders.

Power and Space

To inhabit territory is also to control it, which requires the consolidation of power. To acquire more territory involves the mobilization of the coercive base of power for actual or potential external conflict with other populations inhabiting or making claims to territories. The success of such mobilization depends upon a population's size (for standing armies and domestic labor) and its capacity to produce sufficient material surplus to sustain coercive forces for extended periods of time in the face of resistance. Other factors, such as marchland advantage, increase the likelihood of successful mobilizations, but with continued success this advantage is reduced and continued mobilization depends upon the ability to keep large numbers of coercive forces in the field.

The consolidation of power (PO_{CL}) is essential to bringing together the administrative, material, and legitimating symbols necessary to organize an effective coercive force; and as long as success ensues, legitimating symbols reinforce the distortions in the allocation of material surplus and administrative structures for coercive purposes. As coercion is pursued, centralization of power (PO_{CT}) is ever more necessary to coordinate and control administrative-coercive activity in ever more remote territories. Such centralization and the distortions and limitations on population members, production, and distribution that it creates will be tolerated as long as external conflicts are successful; when they fail, delegitimation rapidly occurs, often to the point of internal conflict that causes disintegration, for a time, of a population (see equation [10] for more discussion).

The continued control of a territory depends upon the same conditions that initially caused annexation of additional space. Initially, such control must be coercive, but if success is achieved over a long time and if assimilation and extensive migration to and from the old and new territories can occur, then coercion can be relaxed. Control of territory now rests on the potential capacity to mobilize defensive forces, but without the excessive distortions of material resources, administrative structures, and legitimating symbols that come with active external conflict. Yet demobilization can prove difficult, setting a coercive regime on the path of overregulation to the point of disintegration.

Law 7: The Density of Settlements in Space

In chapter 6, I viewed density in terms of the overall level of agglomeration of a population in dense settlements. Let me simply rephrase this argument in terms of overall density in the use of space, or (TS_D). Equation (7) presents, in broad strokes, the basic forces involved.

$$(TS_D) = [(P) \times (DF_{MS,MY}) \times (PO_{CL,CT}) \times (N)] + (-/{=}TS) \quad (7)$$

where:

TS_D = the proportion of the population in a territorial space (TS) living in dense settlements ($_D$)

P = the level of production

$DF_{MS,MY}$ = the level of differentiation (DF) of market systems ($_{MS}$) and generalized money media ($_{MY}$)

$PO_{CL,CT}$ = the level of consolidated ($_{CL}$) power (PO) and the degree of centralization ($_{CT}$) of such consolidated power

N = the size of a population

TS = the amount of territorial space inhabited and controlled by a population

Equation (7) simply states that the overall level of density in settlement patterns of a population is a positive function of the multiplicative relations among production (P), market differentiation and use of money $(DF_{MS,MY})$, consolidation and centralization of power $(PO_{CL,CT})$, and population size (N), while bearing a negative relation (that levels off) to the size of territory.

Production and Population Density

As production increases, the level of physical and human capital becomes more concentrated, although this relationship does not kick in until relatively high levels of production are evident. Moreover, as production increases, communication and transportation technologies, as well as the physical infrastructures they generate, also expand and enable movement of people, goods, and materials from the countryside to dense settlements—a trend which, once initiated, leads to larger and more dense settlements. But much of the effect of production on spatial density works through its influence on market and power dynamics.

Markets and Population Density

Centers of trade and exchange pull populations to urban settlements, initially only temporarily. But, as markets expand and use generalized media of exchange, they create resource niches for growing numbers and varieties of specialties and, thereby, increase the material and administrative infrastructures as new organizational units settle near centers of exchange. And because high volume and velocity markets generate increasing wealth, they help finance other nonmarket structures, such as governmental and religious organizations, which pull more people and organizational units to settlements.

Once markets become the principle by which most exchange transactions are conducted, market forces differentiate settlements in terms of prices for land and structures, forcing less affluent purchasers of land or buildings further and further out. As a result, markets extend the boundaries of settlements and, eventually, create the large urban corridors and mega-cities of the late twentieth century. Such a differentiated urban area creates vast numbers of resource niches that pull more individuals within and, if possible, outside of a society to urban centers.

Power and Density

Power ultimately exists as an administrative-coercive capacity that requires space and structures to house its incumbents. Moreover, the nature of power requires consolidation of its bases, causing the holders of power to settle in, or to create, central locales from which administrative, coercive, and incentive activities are conducted. As such, power creates not only infrastructures; its use also generates resource niches for market activities and for other organizational units involved in the use of power. All of these forces, together, pull people and organizational units to the center of power.

As power becomes more centralized, these trends are accelerated, because the administrative, coercive, and incentive systems are pulled toward the centers of decision making. At the same time that power is centralized in a major core settlement, however, it must also develop regional centers to administer and enforce its decisions; and as a result, power creates new dense settlements, or accelerates the urbanization of older ones that may have originally arisen from productive, market, or religious forces. The end result is an expanded infrastructure and a set of new resource niches that pull people and organizational units to secondary centers of power.

Population and Density

A larger population either becomes more densely settled or begins to seek additional territory. When the acquisition of new territories is not possible because of geographical, ecological, or geopolitical constraints, the size and density of settlements increase. And, as this process continues, the needs of the urbanizing population create resource niches for infrastructural development, market activities, production, and regulation by governmental offices. These, in turn, generate further resource niches that pull more people and organizational units into the emerging urban center(s).

Size of Territory and Density

The effect of size of territory is, initially, to decrease settlement patterns, especially if a population is small relative to the land mass of its territory and if production, markets, and power-use are not extensive. Under these conditions, there is little necessity to create dense settlements. But this negative effect vanishes, or perhaps becomes a positive effect, when a population grows, expands production, extends markets, or mobilizes power. For these forces increase density of settlements, and size of territory can cause their mobilization along several fronts: a low ratio of population to land mass reduces the bioecological constraints on population growth; production will expand with population growth; markets will be extended with increased production but also with territorial size as regionally based goods and services are exchanged; and power is mobilized with population growth, with internal or external threats, and with the material surplus created by expanded production and extended markets. And once these forces generate increasingly dense settlements, the infrastructure and resource niches of these settlements pull more people and organizational units into the emerging urban core(s).

Law 8: Structural Differentiation

Differentiation operates at two related levels: structural and cultural. At the structural level (DF_{ST}), analysis focuses on the forces generating diverse positions lodged in different types of organizational units, whereas at the cultural level (DF_C), emphasis is on the forces producing distinctive types of symbol systems and categoric units. Here I examine structural differentiation. Law 9 will be on cultural differentiation.

In equation (8), the forces causing the differentiation of positions and corporate units organizing a population are summarized as an elementary principle:

$$DF_{ST} = [(N) \times (P) \times (DF_{MS,MY}) \times (PO_{CL})] + (+/-PO_{CT}) + (CA_{TR,CM}) + (VS_{GN}) \quad (8)$$

where:

DF_{ST} = the degree of differentiation (DF) among positions and structural units ($_{ST}$) organizing members of a population

N = the size of a population

P = the level of production

$DF_{MS,MY}$ = the level of differentiation (DF) in market systems ($_{MS}$) and generalized money media ($_{MY}$)

PO_{CL} = the degree of political (PO) consolidation ($_{CL}$)

PO_{CT} = the degree to which power (PO) is centralized ($_{CT}$)

$CA_{TR,CM}$ = the level of capital development (CA) of the transportation ($_{TR}$) and communication ($_{CM}$) systems

S_{GN} = the level of generality ($_{GN}$) in value system (VS) premises

Equation (8) argues that the level of structural differentiation (DF_{ST}) evident for a population is: a positive function of the multiplicative relations among population size (N), production (P), differentiation of market systems using money as their media of exchange ($DF_{MS,MY}$), and consolidation of power (PO_{CL}); a positive curvilinear function of political centralization (PO_{CT}); a positive function of the level of capital development in transportation and communication infrastructures

($CA_{TR,CM}$); and a positive function of the level of generality in value system premises (VS_{GN}).

Population and Structural Differentiation

A larger population generates, as Spencer argued, selection pressures for new kinds of productive, distributive, and regulatory units to sustain, support, coordinate, and control the larger "social mass." Once these new types of productive, regulatory, and distributive units come into existence, they create new resource niches for further differentiation, as was discussed earlier for equation (1). Thus, it is Spencerian selection pressures that set into motion the search by members of a growing population for new productive, distributive, and political arrangements; and as these are put into place, they operate to differentiate a population further.

One additional set of selection pressures is more Durkheimian: If population growth causes an increase in the density of settlements (see discussion for equation [7]), then competition often leads individuals and organizational units to seek, or even to create, new resource niches. The end result is increased differentiation among structures as they search for viable resource niches in which to survive. And the more intense the competition for these niches, the greater will be the rate of differentiation.

Production and Differentiation

To increase levels of production requires specialization of structural positions and the organizations housing these positions. For as long as production depends upon a few positions, each performing a multiple set of tasks, output will be limited, but once tasks are divided and coordinated, production can increase. Thus, any significant increase in production causes differentiation of positions and, eventually, the structural units in which these positions are lodged. And once the expansion of production is fueled by inanimate sources of energy, and by market competition, differentiation accelerates and, perhaps, reaches what George Ritzer has termed "hyper-differentiation" as productive (and distributive) units seek new resource niches.

Markets and Differentiation

The expansion of markets is initially caused by long-distance trade and expanding production, but as markets themselves begin to differentiate in response to the growing volume and velocity of transactions as well as to the use of generalized media, they become the primary engine of differentiation, as I discussed in chapters 4 and 7. They do so by enabling individual needs and tasks to be more readily and rapidly translated into ever more diverse market demand, which, in turn, establishes new niches for organizational units. Moreover, once markets are differentiated, units in them seek to create new demands for goods and services, thereby differentiating resource niches further. And because of the tendency of markets using money and other generalized media to pyramid, often into volatile episodes of speculation, they generate additional resource niches for organizational units.

Political Consolidation and Structural Differentiation

There are limits to the amount of differentiation that can occur unless power is consolidated to coordinate and control relations among diverse units. Once initiated, the consolidation of power escalates the potential for further differentiation because it increases the capacity for integration of differentiated units whose conflicts of interest could erupt into disintegrative conflict.

Moreover, the consolidation of power is itself a major axes of differentiation into diverse units associated with coercion, administration, manipulation of incentives, and maintaining legitimacy. And the more these bases of power are needed to manage differentiation outside of the house of power, the more each base internally differentiates to manage the increased demands for regulation. And the differentiation of power can cause its own integrative problems, requiring new kinds of units to coordinate the functions of government. And because power, once consolidated, is a force that tends to be used and extended, even when unnecessary, the structure of government will tend to grow, differentiate, and create needs for further differentiation to integrate the previous round of growth and differentiation. It is not surprising, therefore, that in highly productive and differentiated populations, the expansion of government is one of the major arenas for new patterns of structural differentiation and organizational proliferation.

Political Centralization and Structural Differentiation

The consolidation of power initiates centralization; and when the conditions specified in equation (5) exist, then centralization of decision making, administration, and coercion accelerates. A certain degree of centralization facilitates structural differentiation, along the same lines enumerated above, but highly centralized power (*a*) will generate disincentives for the technological innovations that fuel the differentiating effects of expanded production (because the outputs will be usurped for narrow political purposes) and (*b*) will stifle market forces which increasingly become the engine of differentiation in highly productive populations (because, once again, the incentives for market speculation are reduced and, additionally, because markets generate, as Vilfredo Pareto emphasized, liberal ideologies that can undermine the legitimacy of highly centralized political regimes).

Thus, through overregulation, if not outright repression, highly centralized power usurps the resources and incentives for those productive and market forces that ultimately fuel structural differentiation outside the political sector. And hence, the relationship of centralized power is, as early scholars such as Herbert Spencer and Vilfredo Pareto recognized, positively curvilinear: initial centralization enables necessary coordination and control to proceed in ways encouraging differentiation, whereas too much centralization causes overregulation of production and markets, thereby lessening the capacity for further differentiation and, in fact, often initiating de-differentiation.

Capital Investment in Infrastructures and Structural Differentiation

Durkheim ([1893] 1933) had the rather vague idea that transportation and communication facilities increased "material density" (through its effects on migration and shortening the "distance" between individuals), which, in turn, caused competition and differentiation. Hawley (1986) has perhaps best represented this process as a "mobility cost" variable, because the costs (in energy, time, and money) of moving people, materials, and information restrict the amount of differentiation that can occur. For differentiation is only possible if diverse units can be connected in terms of the flow of information, the transport of people and materials, and the exchange of resources. Markets facilitate the latter, but there must also be a set of physical structures to move information, goods, and people from place to place and from one structural unit to another.

Thus, the more physically developed and technologically advanced are the systems of transportation and communication, the greater are the volume and speed in the movement of resources among units and, hence, the greater their capacity to differentiate from each other. Moreover, capital investment in transportation and communication infrastructures is itself an axes of differentiation, as Spencer argued when he viewed "distribution" as one of the three main axes of functional differentiation (along with production and regulation). But more is involved: this infrastructure creates new opportunities and potential resource niches into which other types of units in the productive, market, and power sectors of a population can move and, in the process, differentiate additional structural forms.

Value Generalization and Structural Differentiation

As Durkheim ([1893] 1933) first emphasized and as Parsons (1966) later argued, structural differentiation cannot occur without the generalization of value premises into highly abstract tenets. For if values are tied to specific structural forms, their elaboration and differentiation is seen to violate basic moral premises. Although differentiation may cause values to generalize in ways that can incorporate ever more diversity, there is a reverse causal effect: generalized value premises facilitate differentiation of structural units, especially when they can form the premises of more specific normative, contractual, and regulatory procedures for coordinating and controlling relations within and between new types of structural units. For without generalized values, differentiation takes on the quality of a "revolt" to established and highly particularized values.

Law 9: Cultural Differentiation

Culture consists of systems of symbols that members of a population create and use to organize activities. At the macro structural level of analysis, three classes of such symbols are important: (1) those symbols that embody value standards (e.g., religious precepts, constitutional premises, moral codes), (2) those that regulate specific classes of activities within and between social units (e.g., norms, laws), and (3) those that denote categories of individuals or social units (e.g., ethnic, religious, class, regional). Durkheim ([1893] 1933) was the first to recognize fully that the degree of homogeneity vs. heterogeneity within and between these systems of symbols is an important property of human organization; and it is in this spirit that I offer an elementary principle on the level of differentiation (DF_C) *among* and *within* these three classes of symbols.

$$DF_C = [(DF_{ST}) \times (N)] + (TS) + (+/-I) \qquad (9)$$

where:

DF_C = the degree of differentiation (DF) within and among value standards, regulatory norms, and social categories ($_C$)

DF_{ST} = the degree of differentiation (DF) among positions and structural units ($_{ST}$) organizing members of a population

N = the size of a population

TS = the size of the territory inhabited and controlled by a population

I = the degree of inequality in the distribution of valued resources among members of a population

Equation (9) states that the differentiation of culture (DF_C) is positively related to the multiplicative relation between the degree of structural differentiation (DF_{ST}) and population size (N), plus the combined effects of the amount of territorial space inhabited and controlled by a population (TS) and the level of inequality (I).

Structural and Cultural Differentiation

Much of the diversity of cultural symbols is caused by structural differentiation of positions and organizational units. As Durkheim ([1893] 1933) recognized, distinctive structural units create systems of regulatory symbols that are separate from value standards; and as Spencer ([1874–96] 1898) implied, structural differentiation establishes a new bases for designating diverse social categories. Indeed, if cultural differentiation does not ensue with initial structural differentiation, the latter is arrested because of an inability to develop regulatory norms and rules for diverse activities within and between units (in the face of heavy Spencerian pressures for such systems of symbols).

Thus, as Adam Smith ([1776] 1805), Niklas Luhmann (1982), Talcott Parsons (1966, 1971), and others have argued, the failure to institutionalize "positive law" in many differentiating populations has presented a barrier to further differentiation and development. Moreover, as emphasized in equation (8), value standards must also become differentiated from diverse sets of regulatory symbols by their abstractness and generality, which, as Adam Smith and the long French lineage culminating with Durkheim emphasized, enable new types of rules, laws, and procedures to emerge that do not violate moral codes. Similarly, once value standards are generalized, and accompanied by differentiation of regulatory symbols, new social categories can be denoted in ways that do not violate values and, at the same time, reflect not only the increasing diversity of structures organizing a population but also the differences among subpopulations in terms of their position inside or outside these structures, their shares of resources, their location in territories and settlements, and their cultural backgrounds. And especially important, with value generalization caused by structural differentiation, new generalized media of exchange can develop and be used to conduct transactions within and between structural units.

Population and Cultural Differentiation

As populations grow, the density of ties among members and the units organizing these members decreases, thereby enabling subpopulations to develop distinctive subcultures and organizational patterns. At a minimum, these events create distinctive categories of individuals, but typically, these categories also develop their own value standards and regulatory norms. These tendencies are accelerated by spatial distributions in territories and settlements of subpopulations and social categories, by patterns of affiliation in particular sets of differentiated structures, and by resource shares in a system of inequality, but population growth, per se, represents an inherent force behind cultural differentiation. And if population growth is driven by immigration, cultural differentiation of value standards, regulatory norms, and social categories will occur, although this process can be reduced over time if assimilation is a viable option and opportunity for immigrants.

Territory and Cultural Differentiation

As Spencer ([1862] 1898, [1874–96] 1898) emphasized, if subpopulations are located in diverse environments, they develop distinctive characteristics, and increasingly become differentiated from each other. As the size of territory increases, then, it is more likely that members of a population will organize and develop cultural systems adapted to diverse circumstances. High rates of mobility along well-developed transportation and communication infrastructures can reduce this tendency for regional subcultures to form, with their own distinct value standards, regulatory norms, and categoric units, but even under these conditions, subcultures can be difficult to break down if they are associated with long-term control of territories and unique organizational patterns. As the collapse of Yugoslavia, the disintegration of the old Soviet Empire, the recent partitioning between the Czech and Slovak Republics, and the continued salience of old tribal conflicts in Africa all document, cultural differences that are tied to territories are extremely difficult to eliminate. Thus, territorial expanse and patterns of settlement are a powerful force behind cultural differentiation.

Inequality and Cultural Differentiation

When resources are unequally distributed, those with different shares will form "class" subcultures. If other subcultural forces, such as ethnic or regional, are also associated with inequality, these cultural differences will be even greater. For variations in resource shares become markers of social categories and what Hawley (1986) termed categoric units; and out of such categories can also emerge distinctive value standards and regulatory symbols, especially when network density of those in a category is high and when distinctive sets of organizational structures (e.g., religious, economic, or familial) correspond to categoric unit membership.

This relationship may, however, be somewhat curvilinear because very high degrees of inequality tend to reduce the number of categoric units to the "haves" and "have-nots." But even here, if other categoric distinctions by ethnicity, religion, region, and origins are associated with a polarized system of resource distribution, competition among the "have-nots" of these additional categories can be highly intense and sustain high levels of cultural differentiation.

Law 10: The Disintegration of Populations

When disintegrative pressures increase, the capacity of a population to remain a coherent whole in its environment is reduced. All known populations have disintegrated in this sense: a particular pattern of societal social organization in space has broken down for a time. Often, new integrative forces reassert themselves, enabling a people to retain elements of cultural and organizational systems across episodes of disintegration. And yet history is the graveyard of most societies that have ever existed; and there is no reason to assume that the current profile of societies in the world will remain. Indeed, the force of disintegration is everywhere in the world today, and even relatively coherent populations face powerful disintegrative pressures. Thus, it is appropriate that this series of elementary principles of macrodynamics close with one on the fate of all organized populations.

$$DS = \overset{(1)}{[(N) \times (TS) \times (-/+DF_S) \times (-/+DF_{C)})]} - \overset{(2)}{[(DF_{MY}) \times (MO)} \times (CA_{TR,CM})] + \overset{(3)}{[(-/+PO_{CT}) \times (I) \times (TH_{IN,EX})]} \quad (10)$$

where:

- DS = the level of disintegrative potential, or actual rate and degree of disintegration of a population, so as to constitute a less coherent and viable whole in its environment
- N = the size of a population
- TS = the size of the territorial space inhabited and controlled by a population
- DF_C = the degree of differentiation (DF) among and within value standards, regulatory norms, and social categories ($_C$)
- DF_{ST} = the level of differentiation (DF) among positions and structural units ($_{ST}$) organizing members of a population
- DF_{MY} = the level of differentiation (DF) of money ($_{MY}$) as a medium of exchange
- MO = the rates of spatial and positional mobility among members of a population
- $CA_{TR,CM}$ = the level of capital investment (CA) in transportation ($_{TR}$) and communication ($_{CM}$) infrastructures
- PO_{CT} = the degree to which power (PO) is centralized ($_{CT}$)

I = the level of inequality in the distribution of valued resources among members of a population

$TH_{IN,EX}$ = the level of internal ($_{IN}$) and external ($_{EX}$) threat (TH) perceived by members of a population and/or its key decision makers in centers of power

Equation (10) is more complicated than those presented earlier because it incorporates many of the forces discussed in the previous nine equations. But in fact, the argument is fairly simple. The disintegrative forces evident for a population can be visualized in three major blocks of forces, marked off by brackets connected by a positive or negative sign, and numbered at the top; the forces within each of the three blocks have multiplicative effects on each other as they increase or decrease the disintegrative pressures on a population.

In the first block, the size of a population (N), the size of its territory (TS), the differentiation of structure (DF_S), and the differentiation of culture (DF_C) affect each other as they work to increase disintegrative potential (DS). Size of the population and the amount of territory controlled by a population exhibit linear positive effects on disintegration, whereas cultural and structural differentiation are negatively curvilinear, working at first to decrease disintegrative potential but then operating to increase such potential as they reach high values.

In the second block are forces that counteract those in the first block: differentiation of money and other generalized media (DF_{MY}), high rates of spatial and positional mobility (MO), and capital investments in communication and transportation infrastructures ($CA_{TR,CM}$) all counter the disintegrative effects of population size, territorial expanse, and sociocultural differentiation.

In the third block are those forces that accelerate and intensify the disintegrative pressures of the first block of forces and, in the end, overcome the mitigating effects of those in the second block: the centralization of power (PO_{CT}), which is negatively curvilinear in its effects on disintegration (decreasing disintegrative pressures at first and, then, increasing them as centralization reaches high levels), inequality over the distribution of resources (I), and internal as well as external threat ($TH_{IN,EX}$). All of these are multiplicatively related, accelerating each other as they affect disintegration. Let me now describe each relation between these forces and disintegration.

Block (1) Disintegrative Forces: Size, Space, and Differentiation

Population Size and Disintegration

As I have stressed, the absolute size and rate of growth in a population increase both logistical loads and potential for a Malthusian correction. As more people must be biologically sustained and coordinated, Spencerian selection pressures for new kinds of structures and cultural symbols escalate, but often fail to create the integrative structures and symbols necessary to support and regulate the increased "social mass," to borrow Herbert Spencer's phrase. As a consequence, disintegration ensues.

Territory and Disintegration

As Spencer ([1874–96] 1898) recognized, the size of a population's territory has two types of disintegrative effects: (1) isolation of subpopulations in local environments where they develop unique and often incompatible patterns of organization, and (2) escalation of logistical loads for transportation, communication, coordination, and control. When territorial size has increased by conquest and annexation, these two disintegrative processes intensify because the conquered will reveal distinctive and often incompatible sociocultural patterns and because coordination and control of restive subjects dramatically raises the level of logistical loads for social control.

Structural Differentiation and Disintegration

The differentiation of structural positions, organizational units, and resource niches all operate to decrease disintegrative pressures, especially under the logistical loads created by population growth. But this relationship between structural differentiation and disintegrative pressures is negatively curvilinear in that high degrees of differentiation increase logistical loads for coordination and control, often to the point of overloading existing distributive and regulatory structures. If lines of structural differentiation are consolidated with patterns of inequality, categoric group formation, or spatial distribution, this disintegrative potential is considerably greater because the force of differentiation is amplified by especially powerful disintegrative forces.

Cultural Differentiation and Disintegration

Like structural differentiation, the differentiation of cultural symbols, especially among abstract value premises, beliefs, and regulatory norms, promotes integration. But high degrees of cultural differentiation begin to pull a population apart and divide its members into particularistic subcultures and social categories whose members have ever fewer symbols in common and, in fact, whose members and organizational units may come into conflict over their respective symbols. Patterns of conquest and annexation, migration, inequality, and spatial isolation all amplify this effect of cultural differentiation and, eventually, drive a population apart.

As the multiplication signs among this first block of forces indicate, population size, amount of territory, and sociocultural differentiation exert their effects on disintegration through their interrelationships with each other. Because of this, they can unravel a population very rapidly once population size and territory create high logistical loads and once the initially integrating effects of cultural and structural differentiation are overcome. For a time sociocultural differentiation revolves around dealing with the logistical loads of increasing population size and territorial expanse, but when such differentiation itself poses integrative problems, the disintegrative effects of all working together can be dramatic, and a seemingly integrated population can unravel very fast.

Block (2) Integrative Forces: Money, Mobility, and Infrastructure

Counteracting the disintegrative effects of block (1) forces are those that arise out of the selection pressures generated by these block (1) forces: differentiation of money as a generalized medium of exchange (DF_{MY}), high rates of vertical and horizontal mobility (MO), and capital investment in transportation and communication infrastructures ($CA_{TR,CM}$). These forces mitigate against the effects of population size, territorial expanse, and differentiation by providing new mechanisms for contact, communication, coordination, and regulation.

Money and Disintegration

Money emerges as a generalized medium of exchange in expanding markets; and although money can exert disintegrative effects when it is implicated in overspeculation in pyramiding metamarkets, it also counteracts the effects of other disintegrative forces. For as I emphasized in chapter 4, money breaks down particularism by virtue of (*a*) its capacity to be used across structures, categories, and subcultures as a neutral unit of value and (*b*) its capacity to bestow value in diverse situations. Moreover, because money facilitates exchange transactions, it increases the volume, velocity, and reach of transactions that actors can have, thereby reducing some of the partitioning effects of differentiation and territorial expanse. Money also becomes an important symbol of collective identification and unification (e.g., the dollar is a symbol of American culture); and when stable, money becomes a potent force behind political legitimation and regulatory power and, hence, an indirect force for integration via its effects on power. Thus, although money is often portrayed as a "root of all evil" and other such things, it is in fact a crucial force in counteracting the disintegrative effects of other forces, a fact that only Georg Simmel ([1907] 1978) among the early sociologists seems to have fully recognized.

Mobility and Disintegration

Movement in space, across categories and organizational units as well as up or down resource hierarchies fosters the development of cross-cutting affiliations and cosmopolitan orientations, while reducing the salience of particularistic symbols. As such, it mitigates against the partitioning effects of sociocultural differentiation and encourages the development of abstract value standards, the articulation of regulatory symbols to coordinate exchanges, and the entrance into ever more places, positions, and categories using money as the "ticket" and "credential" necessary for incumbency and participation.

Capital Investment and Disintegration

As Hawley (1986) recognized, the mobility costs represent crucial forces in the capacity to differentiate in the first place and, then, in the ability to integrate diverse corporate and categoric units created by such differentiation. Thus, reducing the time, energy, and money involved in moving people, materials, resources, and information about a population and its territory is, by itself, a powerful integrative force. Although reducing mobility costs is dependent on technologies, the really important consideration is actual capital investments in communication and transportation infrastructures. Technology without its implementation as physical facilities has little effect, but when the physical capital exists, its very presence creates bridges across territories and ties across groups or categories. And independent of their effects on mobility and market transactions using money, which have the effects of reducing partitions and particularism discussed above, the existence of communication and transportation infrastructures becomes a cultural symbol—indeed, a kind of totem—signifying that access to other places and positions is possible and, as such, fosters more cosmopolitan and less particularistic orientations. Moreover, this infrastructure can often serve as a symbol of the effectiveness of power, thereby providing a source of political legitimation. And so, societies without well-developed communication and transportation infrastructures reduce the ties and bridges connecting members and the units organizing and categorizing these members, thereby encouraging divisions, partitions, and particularisms that inevitably increase pressures for disintegration.

These forces in this second block are multiplicatively related in terms of their effects on disintegration. Each amplifies the others in ways that create a set of counterforces to those denoted in block (1) and, as we will see, block (3). Indeed, while rather undertheorized in sociology, it could be argued that the forces in block (2), along with the initial integrative effects of sociocultural differentiation and centralization of power, are what really hold a larger population in an extended territory together, at least for a time. And when sociocultural differentiation and centralization begin to shift toward the disintegrative side, it is these forces in block (2) that sustain a population's coherence in its environment, even under the mounting disintegrative pressures generated by population size, extended territory, sociocultural differentiation, and centralized power as it affects inequality and mounting threats.

Block (3) Forces of Disintegration: Power, Inequality, and Threat

The most volatile set of forces behind disintegration is the accelerating effects among the centralization of power, growing inequality, and escalating internal threats accompanied by external threats. When block (1) forces are coupled with high values for these block (3) forces, then the mitigating effects of block (2) forces are eventually overwhelmed, causing a society to disintegrate rapidly.

Power and Disintegration

The consolidation of the four bases of power initiates the process of centralization of power; further centralization is often necessary to regulate, coordinate, and control activities. But whether from its own inertia (i.e., power begets efforts to secure more power) or from the escalating effects among rising inequality, internal threats, and external threats, a high degree of centralized power is as much disintegrative as integrative. As I emphasized in chapters 5 and 8, overregulation generates resentments and the mobilization of counterpower, and especially so if accompanied by high levels of inequality and rigid stratification. As internal threats mount from the mobilization of counterpower, centralization is increased in order to maintain control; and as this process of centralization and overregulation proceeds, the balance among the four bases of power is distorted toward the administrative-coercive components and away from the co-optive use of material incentives and the nurturing of consensus over legitimating symbols.

Often under these circumstances, external threats increase as other populations see the vulnerability of a political regime; and equally often, external threats are manufactured in order to mobilize a mantra of legitimating symbols and to deflect attention away from rising inequality, rigid stratification, and internal threats. In either case, but particularly if external threats have been manufactured, failure of a regime in the geopolitical arena rapidly delegitimates centers of power and causes the mobilization of the counterpower. Thus, a high degree of centralized power sets into motion volatile disintegrative forces, each of which is examined below.

Inequality and Disintegration

A certain level of inequality and stratification (i.e., formation of "class" categoric units in terms of resource shares) is inevitable as production, power, and population all increase. The disintegrative effects of such stratification are greatly reduced if rates of mobility are high, if boundaries between class categories are ambiguous, and if positions on various resource hierarchies are not highly correlated. When power becomes highly centralized, however, inequality will increase and these mitigating forces will be increasingly eliminated as those with power usurp resources to sustain their privilege. Similarly, if economic cartels dominate production and distribution, inequality will increase, and if such cartels rely upon power to sustain their advantage, then once again the mitigating effects of mobility, ambiguous class boundaries, and unconsolidated positions on resource hierarchies will be reduced.

The existence of high levels of inequality in resource distribution will, inevitably and inexorably, cause stratification to become more rigid, setting into motion the mobilization of counterpower, which typically results in open conflict and at least partial disintegration but which, at times, leads to political reform and redistribution of resources in ways to reduce the level of inequality and the rigidity of stratification (e.g., as the history of most Western democracies illustrates). This latter turn of events is only likely, I argue, under conditions of increasing differentiation of markets and widespread use of money ($DF_{MS,MY}$). If ($DF_{MS,MY}$) is not in place, then the mobilization of counterpower will either fail and cause further centralization of power to reassert repressive control or it will be successful and create yet another center of repressive power to guard against counterrevolutionary forces. The repressiveness of the aftermath of the Russian Revolution was, I think, an example of revolt against high levels of inequality and stratification without high levels of differentiation of markets and widespread use of generalized media like money. The results were predictable: increasing inequality, stratification, and internal threat leading to more centralized power, which only locks a population into an escalating cycle of increasing (PO_{CT}), (I), and TH_{IN}).

Threat and Disintegration

Threat is the general label that I employ to denote the mobilization of counterpower which challenges existing centers of power. Such threats can be internal ($_{IN}$) or external ($_{EX}$), and despite differences in the dynamics unleashed by these two sources of threat, they both challenge the legitimacy and prerogatives of existing centers of power to regulate, coordinate, and control a population in certain ways.

As threats increase, the probability for conflict also rises, causing existing centers of power to mobilize symbols, material resources, administrative capacities, and coercive forces to deal with the threat. As argued above, however, such mobilization and centralization of power set into motion those very processes that escalate internal threat and make external threats increasingly likely to emerge. Thus, once high levels of threat exist and mobilization of internal counterpower or external power is under way, disintegration becomes increasingly likely.

A Final Word

In closing, let me repeat that the word "toward" in the subtitle of this book should be taken seriously. I am working toward a general theory; as my critics will no doubt point out, I am not there yet. For some, I will never get there with my naive positivists' assumptions about how to develop theory.

But if one does not like where I take the basic ideas—which are, in reality, a synthesis of well-known approaches—then I invite others to go their way; and then, let us compare notes. And so, I offer this appendix as a starting point for engagement with my fellow theorists—or, at least those who think it possible to explain the operative dynamics of the universe—and with researchers who seek to use data of whatever source to illustrate underlying and generic social processes.

NOTES

4. Distribution Dynamics

1. I will emphasize market, and market differentiation, as the key exchange force, although I perhaps should remain more abstract. For markets may not exist centuries into the future; there will, no doubt, be exchange distribution in the future, but market distribution may be a historically time-bound phenomenon. Still, on the other hand, markets may be more generic than I am implying here, and so this emphasis on markets may not be wholly inappropriate.

5. Power Dynamics

1. For example, Emerson 1972; Willer 1986; Fararo 1989; Blalock 1989; Wrong 1979; Mann 1986; Moore 1966; Haas 1982; Mair 1962; Blau 1964; Homans 1961; Fried 1967; Collins 1975; Rueschemeyer 1977, 1982; Skocpol 1979; Luhmann 1982; and so on for many more scholars.

6. Spatial Dynamics

1. A more detailed review of urban ecology as it has affected my thinking here can be found in Jonathan H. Turner, "The Assembling of Human Populations," *Advances in Human Ecology* 3 (1994): 65–91.
2. For a review of these as they relate to my argument here, see my "Assembling of Human Populations," *Advances in Human Ecology* 3 (1994): 65–91.

7. Differentiating Dynamics

1. In recent years, there have been numerous extensions of the synthetic theory of evolution to sociological work, including sociobiology (e.g., van den Berghe 1978;

Lopreato 1984), dual inheritance (e.g., Boyd and Richerson 1985), and coevolution (e.g., Durham 1991).

2. For example, Parsons (1966); Alexander (1985); Alexander and Colomy (1990); and Colomy (1990).

3. Park and Burgess (1925); Burgess (1925); McKenzie (1933); Park (1936); Hawley (1971/1981); Berry and Kasarda (1977); Kasarda (1972); Castells (1985, 1988); Frisbie (1980); Frisbie and Kasarda (1988).

4. Hannan and Freeman (1977, 1984, 1986, 1987, 1988, 1989); McPherson (1981, 1983a, 1983b, 1988, 1990); Bidwell and Kasarda (1985, 1987).

5. A more detailed presentation of many ideas in this chapter was originally published in my article "The Ecology of Macrostructure" in *Advances in Human Ecology* 3 (1994): 113–37.

6. This model is a composite constructed from a number of works: Hannan and Freeman (1977, 1984, 1986, 1987, 1988, 1989); McPherson (1981, 1983a, 1983b, 1988, 1990); McPherson and Ranger-Moore (1991); McPherson and Smith-Lovin (1988); McPherson, Popielarz and Drobnic (1992); Bidwell and Kasarda (1985, 1987); Aldrich (1979).

7. This model is constructed from Hawley (1944, 1950, 1973, 1978, 1981, 1986).

REFERENCES

Aldrich, Howard. 1979. *Organizations and Environments.* Englewood Cliffs, N.J.: Prentice-Hall.

Alexander, Jeffrey C. 1985. *Neofunctionalism.* Newbury Park, Calif.: Sage.

Alexander, Jeffrey C., Bernhard Giesen, Richard Münch, and Neil J. Smelser. 1986. *The Micro-Macro Link.* Berkeley and Los Angeles: University of California Press.

Alexander, Jeffrey C., and Paul Colomy, eds. 1990. *Differentiation Theory and Social Change.* New York: Columbia University Press.

Appelbaum, Richard P. 1978. "Marx's Theory of the Falling Rate of Profit; Toward a Dialectical Analysis of Structural Change." *American Sociological Review* 43:64–73.

Baran, Paul, and Paul M. Sweezy. 1966. *Monopoly Capital.* New York: Monthly Review Press.

Bender, Barbara. 1975. *Farming in Prehistory.* London: Baker.

Berry, Brian J. L., and John D. Kasarda. 1977. *Contemporary Urban Ecology.* New York: Macmillan.

Bidwell, Charles E., and John D. Kasarda. 1985. *The Organization and Its Ecosystem: A Theory of Structuring in Organizations.* Greenwich, Conn.: JAI Press.

———. 1987. *Structuring in Organizations: Ecosystem Theory Evaluated.* Greenwich, Conn: JAI Press.

Blalock, Hubert M., Jr. 1989. *Power and Conflict: Toward a General Theory.* Newbury Park, Calif.: Sage.

Blau, Judith R. 1993. *Social Contracts and Economic Markets.* New York: Plenum Press.

Blau, Peter M. 1964. *Exchange and Power in Social Life.* New York: Wiley.

———. 1977. *Inequality and Heterogeneity, A Primitive Theory of Social Structure.* New York: Free Press.

Bloch, Marc. 1962. *Feudal Society.* Translated by L. A. Manyon. Chicago: University of Chicago Press.

Block, Fred. 1980. "Beyond Relative Autonomy: State Managers as Historical Subjects." *Socialist Register* 8:227–42.

Boserup, Ester. 1965. *The Conditions of Agricultural Growth.* Chicago: Aldine.

———. 1981. *Population and Technological Change: A Study of Long-Term Trends.* Chicago: University of Chicago Press.

Boyd, Robert, and Peter J. Richerson. 1985. *Culture and the Evolutionary Process.* Chicago: University of Chicago Press.

Braudel, Fernand. 1977. *Afterthoughts on Material Civilization and Capitalism.* Translated by P. M. Ranum. Baltimore: Johns Hopkins University Press.

———. [1979] 1982. *Wheels of Commerce. Civilization and Capitalism, 15th–18th Century.* New York: Harper/Collins.

Burgess, Ernest W. 1925. "The Growth of the City: An Introduction to a Research Project." In *The City,* edited by R. Park, E. Burgess, and R. J. McKenzie, 47–62. Chicago: University of Chicago Press.

Carneiro, Robert L. 1967. "On the Relationship of Size of Population and Complexity of Social Organization." *Southwestern Journal of Anthropology* 23:234–43.

———. 1970. "A Theory of the Origin of the State." *Science* 169:733–38.

———. 1973. "Structure, Function, and Equilibrium in the Evolutionism of Herbert Spencer." *Journal of Anthropological Research* 29 (2):77–95.

Castells, Manuel. 1985. "High Technology, Economic Restructuring and the Urban-Regional Process in the United States." In *High Technology, Space, and Society,* edited by M. Castells, 11–24. Newbury Park, Calif.: Sage.

———. 1988. "High Technology and Urban Dynamics in the United States." In *The Metropolis Era: A World of Giant Cities,* edited by M. Dogan and J. D. Kasarda, 85–110. Newbury Park, Calif.: Sage.

Clark, Colin. 1951. "Urban Population Densities." *Journal of the Royal Statistical Society,* ser. A. 114:490–96.

Coleman, James S. 1986. *Individual Interests and Collective Action.* New York: Cambridge University Press.

———. 1990. *Foundations of Social Theory.* Cambridge, Mass.: Belknap Press.

Collins, Randall. 1975. *Conflict Sociology.* New York: Academic Press.

———. 1981. "On the Micro-foundations of Macro-sociology." *American Journal of Sociology* 86:984–1014.

———. 1986. *Weberian Sociological Theory.* New York: Cambridge University Press.

———. 1988. *Theoretical Sociology.* San Diego: Harcourt Brace Jovanovich.

———. 1990. "Market Dynamics as the Engine of Historical Change." *Sociological Theory* 8:111–35.

Colomy, Paul. 1990. *Neofunctionalist Sociology: Contemporary Statements.* London: Edward Elgar.

Comte, Auguste. 1830–42. *The Course of Positive Philosophy.* London: Bell.

Coser, Lewis A. 1956. *The Functions of Social Conflict.* London: Free Press.

———. 1967. *Continuities in the Study of Social Conflict.* New York: Free Press.

Curwen, Cecil, and Gudmund Hatt. 1961. *Plough and Pasture: The Early History of Farming.* New York: Collier.

Dahrendorf, Ralf. 1959. *Class and Class Conflict in Industrial Society.* Stanford: Stanford University Press.

DiMaggio, Paul J., and Walter W. Powell. 1983. "The Iron Cage Revisited: Institutional Isomorphism and Collective Rationality in Organizations." *American Sociological Review* 48:147–60.

Douglas, Mary. 1966. *Purity and Danger.* London: Routledge.

———. 1973. *Natural Symbols.* Baltimore: Penguin Books.

Durham, William. 1991. *Coevolution.* Stanford, Calif.: Stanford University Press.

Durkheim, Émile. [1893] 1933. *The Division of Labor in Society.* New York: Free Press.

Earle, Timothy, ed. 1984. *On the Evolution of Complex Societies.* Malibu, Calif.: Undena.

Earle, Timothy, and J. Ericson, eds. 1977. *Exchange Systems in Prehistory.* New York: Academic Press.

Elkin, A. P. 1954. *The Australian Aborigines.* 3d ed. Sydney: Angus and Robertson.

Emerson, Richard M. 1962. "Power-Dependence Relations. *American Sociological Review* 17:31–41.

———. 1972. "Exchange Relations and Network Structures." In *Sociological Theories in Progress,* edited by J. Berger, M. Zelditch, and B. Anderson, 38–87. New York: Houghton Mifflin.

Etzioni, Amitai. 1961. *A Comparative Analysis of Complex Organizations.* New York: Free Press.

Fararo, Thomas J. 1989. *The Meaning of General Theoretical Sociology.* New York: Cambridge University Press.

Flannery, Kent V. 1973. "The Origins of Agriculture." *Annual Review of Anthropology* 2:271–310.

Frank, André Gunder. 1979. *Dependent Accumulation.* New York: Monthly Review Press.

Fried, Morton H. 1967. *The Evolution of Political Society.* New York: Random House.

Friedman, William. 1959. *Law in a Changing Society.* Berkeley and Los Angeles: University of California Press.

Frisbie, Parker W. 1980. "Theory and Research in Urban Ecology." In *Sociological Theory and Research: A Critical Approach,* edited by H. M. Blalock Jr, 203–19. New York: Free Press.

Frisbie, Parker W., and John D. Kasarda. 1988. "Spatial Processes." In *Handbook of Sociology,* edited by N. J. Smelser, 629–66. Newbury Park, Calif.: Sage.

Fuchs, Stephan, and Jonathan H. Turner. 1991. "Legal Sociology as General Theory." *Virginia Review of Sociology* 1:165–72.

Goldstone, Jack. 1990. *Revolution and Rebellion in the Early Modern World, 1640–1848.* Berkeley and Los Angeles: University of California Press.

Granovetter, Mark. 1973. "The Strength of Weak Ties." *American Journal of Sociology* 78:1360–80.

Gurvitch, George. 1953. *Sociology of Law.* London: Routledge and Kegan Paul.

Haas, Jonathan. 1982. *The Evolution of the Prehistoric State.* New York: Columbia University Press.

Hall, John A. 1985. *Powers and Liberties: The Causes and Consequences of the West.* Oxford: Blackwell.

Hall, Peter. 1984. *The World of Cities.* 3d ed. London: Weidenfeld and Nicholson.

———. 1988. "Urban Growth and Decline in Western Europe." In *The Metropolis Era: A World of Giant Cities,* 111–27. Newbury Park, Calif.: Sage.

Hannan, Michael T., and John Freeman. 1977. "The Population Ecology of Organizations." *American Journal of Sociology* 82 (March): 929–64.

———. 1984. "Structural Inertia and Organizational Change." *American Sociological Review* 49:149–64.

———. 1986. "Where Do Organizational Forms Come From?" *Sociological Forum* 1:50–72.

———. 1987. "The Ecology of Organizational Founding: American Labor Unions, 1836–1985." *American Journal of Sociology* 92:910–43.

———. 1988. "The Ecology of Organizational Mortality: American Labor Unions, 1836–1985." *American Journal of Sociology* 94:25–52.
———. 1989. *Organizational Ecology.* Cambridge: Harvard University Press.
Hawley, Amos H. 1944. "Ecology and Human Ecology." *Social Forces* 27:398–405.
———. 1950. *Human Ecology.* New York: Ronald Press.
———. [1971] 1981. *Urban Society: An Ecological Approach.* New York: Ronald Press.
———. 1973. "Ecology and Population." *Science* 1979 (March): 1196–1201.
———. 1978. "Cumulative Change in Theory and History." *American Sociological Review* 43:787–97.
———. 1981. "Human Ecology: Persistence and Change." *American Behavioral Scientist* 24:423–44.
———. 1986. *Human Ecology: A Theoretical Essay.* Chicago: University of Chicago Press.
Hechter, Michael. 1987. *Principles of Group Solidarity.* Berkeley and Los Angeles: University of California Press.
Heckathorn, Douglas. 1990. "Collective Sanctions and Compliance Norms: A Formal Theory of Group-Mediated Control." *American Sociological Review* 55 (3):366–84.
Homans, George C. 1961. *Social Behavior: Its Elementary Forms.* New York: Harcourt.
Johnson, Allen W., and Timothy Earle. 1987. *The Evolution of Human Societies: From Foraging Groups to Agrarian State.* Stanford, Calif.: Stanford University Press.
Kasarda, John D. 1972. "The Theory of Ecological Expansion: An Empirical Test." *Social Forces* 51:165–75.
Keohane, Robert O. 1984. *After Hegemony: Cooperation and Discord in the World Political Economy.* Princeton, N.J.: Princeton University Press.
Keohane, Robert O., and Joseph Nye. 1989. *Power and Interdependence.* Glenview, Ill.: Scott, Foresman.
Keohane, Robert O., Joseph F. Nye, and Stanley Hoffman, eds. 1993. *After the Cold War: International Institutions and State Strategies in Europe, 1989–1991.* Cambridge: Harvard University Press.
Klassen, L. H., W. T. H. Molle, and J. H. P. Paelinck, eds. 1981. *Dynamics of Urban Development.* New York: St. Martins Press.
Lee, Richard. 1979. *The ! Kung San.* Cambridge: Cambridge University Press.
Lee, Richard, and Irven DeVore, eds. 1968. *Man the Hunter.* Chicago: Aldine.
———, eds. 1976. *Kalahari Hunter-Gatherers.* Cambridge: Cambridge University Press.
Lee, Ronald Demos. 1986. "Malthus and Boserup: A Dynamic Synthesis." In *The State and Population Theory,* edited by D. Coleman and R. Schofield. New York: Basil Blackwell.
Lenski, Gerhard. 1966. *Power and Privilege.* New York: McGraw-Hill.
Lenski, Gerhard, Jean Lenski, and Patrick Nolan. 1991. *Human Societies: An Introduction to Macrosociology.* New York: McGraw-Hill.
Lloyd, D. 1964. *The Idea of Law.* Baltimore: Penguin Books.
Lopreato, Joseph. 1984. *Human Nature and Biocultural Evolution.* Boston: Allen and Unwin.
Luhmann, Niklas. 1982. *The Differentiation of Society.* New York: Columbia University Press.
———. 1985. *A Sociological Theory of Law.* Boston: Routledge and Kegan Paul.

Mair, Lucy. 1962. *Primitive Government*. Baltimore: Penguin Books.

Malthus, Thomas. [1798] 1926. *First Essay on Population*. New York: Kelley.

Mann, Michael. 1986. *The Social Sources of Power. Volume 1 on: A History of Power from the Beginning to A.D. 1760*. Cambridge: Cambridge University Press.

Marx, Karl. [1867] 1967. *Capital: A Critical Analysis of Capitalist Production*. New York: International Publishers.

Marx, Karl, and Friedrich Engels. [1846] 1947. *The German Ideology*. New York: International Publishers.

———. [1848] 1978. *The Communist Manifesto*. In *The Marx-Engels Reader*, edited by Robert Tucker, 469–500. New York: Norton.

Maryanski, Alexandra, and Jonathan H. Turner. 1992. *The Social Cage: Human Nature and the Evolution of Society*. Stanford, Calif.: Stanford University Press.

McKenzie, Roderick. 1933. *The Metropolitan Community*. New York: McGraw-Hill.

McPherson, J. Miller. 1981. "A Dynamic Model of Voluntary Affiliation." *Social Forces* 59:705–28.

———. 1983a. "An Ecology of Affiliation." *American Sociological Review* 48:519–32.

———. 1983b. "The Size of Voluntary Organizations." *Social Forces* 61:1044–64.

———. 1988. "A Theory of Voluntary Organization." In *Community Organizations*, edited by C. Milofsky, 42–76. New York: Oxford.

———. 1990. "Evolution in Communities of Voluntary Organizations." In *Organizational Evolution*, edited by Jihendra Singh, 224–45. Newbury Park, Calif.: Sage.

McPherson, J. Miller, Pamela A. Popielarz, and Sonja Drobnic. 1992. "Social Networks and Organizational Dynamics." *American Sociological Review* 57:153–70.

McPherson, J. Miller, and J. Ranger-Moore. 1991. "Evolutionary on a Dancing Landscape: Organizations and Networks in Dynamic Blau Space." *Social Forces* 70 (1):19–42.

McPherson, J. Miller, and Lynn Smith-Lovin. 1988. "A Comparative Ecology of Five Nations: Testing a Model of Competition Among Voluntary Organizations." Chap. 6 in *Ecological Models of Organizations*, edited by G. R. Carroll. Cambridge: Ballinger.

Moore, Barrington Jr. 1966. *Social Origins of Dictatorship and Democracy*. Boston: Beacon Press.

Pareto, Vilfredo. [1916] 1935. *Treatise on General Sociology*. Under the title *The Mind and Society*, New York: Harcourt, Brace.

Park, Robert E. 1936. "Human Ecology." *American Journal of Sociology* 42:1–15.

Park, Robert E., and Ernest W. Burgess. 1925. *The City*. Chicago: University of Chicago Press.

Parsons, Talcott. 1951. *The Social System*. New York: Free Press.

———. 1963a. "On the Concept of Political Power." *Proceedings of the American Philosophical Society* 107 (June): 232–62.

———. 1963b. "On the Concept of Influence." *Public Opinion Quarterly* 27 (Spring): 37–62.

———. 1966. *Societies: Evolutionary and Comparative Perspectives*. Englewood Cliffs, N.J.: Prentice-Hall.

———. 1970. "Some Problems of General Theory." In *Theoretical Sociology: Perspectives and Developments*, edited by J. C. McKinney and E. A. Tiryakian. New York: D. Appleton.

———. 1971. *The System of Modern Societies*. Englewood Cliffs, N.J.: Prentice-Hall.

Parsons, Talcott, and Neil J. Smelser. 1956. *Economy and Society.* New York: Free Press.
Radcliffe-Brown, A. R. 1952. *Structure and Function in Primitive Society.* Glencoe, Ill.: Free Press.
Rueschemeyer, Dietrich. 1977. "Structural Differentiation, Efficiency and Power." *American Journal of Sociology* 83 (July): 1–25.
———. 1982. "On Durkheim's Explanation of Division of Labor." *American Journal of Sociology* 88 (November): 579–90.
Sahlins, Marshall. 1972. *Stone Age Economics.* Chicago: Aldine.
Sailer, Lee Douglas. 1978. "Structural Equivalence." *Social Networks* 1:73–90.
Service, Elman. 1962a. *Primitive Social Organization: An Evolutionary Perspective.* New York: Random House.
———. 1962b. *Primitive Social Organization.* New York: Random House.
———. 1966. *The Hunters.* Englewood Cliffs, N.J.: Prentice-Hall.
Simmel, Georg. [1902] 1950. "The Metropolis and Mental Life." In *The Sociology of Georg Simmel,* edited by K. Wolff, 409–24. New York: Free Press.
———. [1907] 1978. *The Philosophy of Money.* Translated by T. Bottomore and D. Frisby. Boston: Routledge and Kegan Paul.
———. 1956. *Conflict and the Web of Group Affiliations.* Translated by K. Wolf. New York: Free Press.
Skocpol, Theda. 1979. *States and Social Revolutions.* New York: Cambridge University Press.
Smith, Adam. [1776] 1805. *An Inquiry into the Nature and Causes of the Wealth of Nations.* London: Davis.
Spencer, Herbert. [1851] 1888. *Social Statics.* New York: D. Appleton.
———. [1862] 1898. *First Principles.* New York: D. Appleton.
———. [1874–96] 1898. *The Principles of Sociology.* 3 vols. New York: D. Appleton.
Stephan, Edward G. 1979a. "Variation in County Size: A Theory of Segmental Growth." *American Sociological Review* 36 (June): 451–61.
———. 1979b. "Derivation of Some Socio-Demographic Regularities from the Theory of Time-Minimization." *Social Forces* 57:812–23.
Stinchcombe, Arthur L. 1968. *Constructing Social Theories.* New York: Harcourt Brace Jovanovich.
Tainter, Joseph. 1988. *The Collapse of Complex Societies.* Cambridge: University of Cambridge Press.
Turner, Jonathan H. 1972. *Patterns of Social Organization: A Survey of Social Institutions.* New York: McGraw-Hill.
———. 1980. "Legal Evolution: An Analytical Model." In *The Sociology of Law,* edited by W. E. Evan, 377–94. New York: Free Press.
———. 1981a. "The Forgotten Giant: Herbert Spencer's Models and Principles." *Revue Europeenne Des Sciences Sociales* 19 (59): 79–98.
———. 1981c. "Émile Durkheim's Theory of Integration in Differentiated Social Systems." *Pacific Sociological Review* 24 (4):187–208.
———. 1981d. "Emile Durkheim's Theory of Integration in Differentiated Social Systems." *Pacific Sociological Review* 24 (4):187–208.
———. 1984a. "Durkheim's and Spencer's Principles of Social Organization." *Sociological Perspectives* 27 (1):21–32.
———. 1984b. *Societal Stratification: A Theoretical Analysis.* New York: Columbia University Press.

———. 1985a. *Herbert Spencer: A Renewed Appreciation*. Newbury Park, Calif.: Sage.

———. 1985b. "In Defense of Positivism." *Sociological Theory* 3 (2):24–30.

———. 1990a. "A Theory of Macrostructural Dynamics." In *Sociological Theories in Progress: New Formulations,* edited by M. Zelditch and J. Berger, 185–214. Newbury Park, Calif.: Sage.

———. 1990b. "The Misuse and Use of Metatheory." *Sociological Forum* 5:37–53.

———. 1990c. "The Promise of Positivism." In *Postmodernism and Social Theory.* New York: Basil Blackwell.

———. 1990d. "Durkheim's Theory of Social Organization." *Social Forces* 68:1–15.

———. 1990e. *The Structure of Sociological Theory.* Belmont, Calif.: Wadsworth.

———. 1991. "Simmel and Weber on Money, Exchange, and Structural Differentiation." *Simmel Newsletter* 1 (2):80–90.

———. 1994a. "The Assembling of Human Populations." *Advances in Human Ecology* 3:65–91.

———. 1994b. "The Ecology of Macrostructure." *Advances in Human Ecology* 3:113–37.

Turner, Jonathan H., and Alexandra Maryanski. 1979. *Functionalism.* Menlo Park: Benjamin-Cummings.

Turner, Jonathan H., and Charles Starnes. 1976. *Inequality: Privilege and Poverty in America.* Santa Monica: Goodyear.

Vaco, Steven. 1988. *Law and Society.* 2d ed. Englewood Cliffs, N.J.: Prentice-Hall.

van den Berghe, Pierre. 1978. "Bridging the Paradigms." *Society* 15 (1978): 42–49.

Wallerstein, Immanuel M. 1974. *The Modern World-System: Capitalist Agriculture and the Origins of the European World Economy in the Sixteenth Century.* New York: Academic Press.

Weber, Max. [1922] 1978. *Economy and Society: An Outline of Interpretive Sociology.* Edited by G. Roth and C. Wittich. Berkeley and Los Angeles: University of California Press.

Webster, D. 1975. "Warfare and the Evolution of the State." *American Antiquity* 40:467–70.

White, Harrison C. 1970. *Chains of Opportunity: System Models in Organizations.* Cambridge: Harvard University Press.

Willer, David. 1986. *Theory and the Experimental Investigation of Social Structures.* New York: Gordon and Breach.

Wrong, Dennis. 1979. *Power: Its Forms, Bases, and Uses.* New York: Harper and Row.

Zipf, George. 1949. *Human Behavior and the Principle of Least Effort.* Reading, Mass.: Addison-Wesley.

NAME INDEX

SUBJECT INDEX

ABOUT THE AUTHOR

Jonathan H. Turner has been a professor of sociology at the University of California at Riverside since 1969. He is the author of two dozen books on sociological theory, social stratification, ethnicity, and American society and social institutions. Past president of the Pacific Sociological Association, he now edits this association's journal, *Sociological Perspectives.* He has served in a variety of additional editorial positions as a member of the board of the University of California Press and Rose Book Series, while editing three separate series of books for publishers. He has also served on the editorial board of numerous journals, including the *American Journal of Sociology,* the *American Sociological Review,* and the *Sociological Quarterly.* While his research has branched into many substantive areas, his core agenda has always been the development of general scientific theory in sociology.